The New Evolutionary Paradigm

Originally published in 1991, *The New Evolutionary Paradigm* provides an innovative and cross disciplinary look at evolution. While Darwin's theory of evolution was originally restricted to the life sciences, the same principles have been applied successfully to historical, social and natural sciences. The papers included in *The New Evolutionary Paradigm* analyse the facts, observations, and accumulated data from the significance of a general evolution theory cannot be overemphasised; a new understanding of the cosmos and man's relationship to it could lead to the systemization of the irreversible change that takes place in society and nature. This book will appeal to scientists, sociologists and those interested in transdisciplinary evolution theories.

T0173814

The New Evolutionary Paradigm

Keynote Volume

Edited by Ervin Laszlo

First published in 1991
by Gordon and Breach Science Publishers

This edition first published in 2019 by Routledge
2 Park Square, Milton Park, Abingdon, Oxon, OX14 4RN
and by Routledge
52 Vanderbilt Avenue, New York, NY 10017

Routledge is an imprint of the Taylor & Francis Group, an informa business

Publisher's Note
The publisher has gone to great lengths to ensure the quality of this reprint but points out that some imperfections in the original copies may be apparent.

Disclaimer
The publisher has made every effort to trace copyright holders and welcomes correspondence from those they have been unable to contact.

A Library of Congress record exists under LCCN: 90041157

ISBN 13: 978-0-367-33921-0 (hbk)
ISBN 13: 978-0-429-32285-3 (ebk)
ISBN 13: 978-0-367-33922-7 (pbk)

The New Evolutionary Paradigm

Keynote Volume

Edited by Ervin Laszlo

First published in 1991
by Gordon and Breach Science Publishers

This edition first published in 2019 by Routledge
2 Park Square, Milton Park, Abingdon, Oxon, OX14 4RN
and by Routledge
52 Vanderbilt Avenue, New York, NY 10017

Routledge is an imprint of the Taylor & Francis Group, an informa business

Publisher's Note
The publisher has gone to great lengths to ensure the quality of this reprint but points
out that some imperfections in the original copies may be apparent.

Disclaimer
The publisher has made every effort to trace copyright holders and welcomes
correspondence from those they have been unable to contact.

A Library of Congress record exists under LCCN: 90041157

ISBN 13: 978-0-367-33921-0 (hbk)
ISBN 13: 978-0-429-32285-3 (ebk)
ISBN 13: 978-0-367-33922-7 (pbk)

THE NEW EVOLUTIONARY PARADIGM

THE NEW EVOLUTIONARY PARADIGM

KEYNOTE VOLUME

Edited by

Ervin Laszlo
The Vienna Academy for the Study of the Future
Austria

Foreword by Ilya Prigogine

Gordon and Breach Science Publishers

New York   Philadelphia   London   Paris   Montreux   Tokyo   Melbourne

Gordon and Breach Science Publishers

Post Office Box 786
Cooper Station
New York, New York 10276
United States of America

5301 Tacony Street, Drawer 330
Philadelphia, Pennsylvania 19137
United States of America

Post Office Box 197
London WC2E 9PX
United Kingdom

58, rue Lhomond
75005 Paris
France

Post Office Box 161
1820 Montreux 2
Switzerland

3–14–9, Okubo
Shinjuku-ku, Tokyo 169
Japan

Private Bag 8
Camberwell, Victoria 3124
Australia

Library of Congress Cataloging-in-Publication Data

The new evolutionary paradigm / edited by Ervin Laszlo.
     p.   cm. — (The World futures general evolution studies :
  v. 2)
   Includes index.
   ISBN 2-88124-375-4
   1. Evolution. I. Laszlo, Ervin, 1932– . II. Series.
QH366.2.N48 1991
575.01—dc20
                                          90–41157
                                             CIP
                                           Rev.

To Jonas Salk, Frederico Mayor and Ilya Prigogine,
and my colleagues in the
General Evolution Research Group

Ralph Abraham
Prince Alfred of Liechtenstein
Peter Allen
Robert Artigiani
Bela Banathy
Thomas Bernold
Gianluca Bocchi
Miriam Campanella
Mauro Ceruti
Erich Chaisson
Allan Combs
John Corliss
Vilmos Csányi
Riane Eisler
György Kampis
David Loye
Pentti Malaska
Edward Markarian
Ignazio Masulli
Jiayin Min
Mika Pantzar
Gerlind Rurik
Singa Sandelin
Peter Saunders
Jonathan Schull
Rudolf Treumann
Francisco Varela

CONTENTS

PREFACE TO THE SERIES

The World Futures General Evolution Studies series is associated with the journal *World Futures: The Journal of General Evolution*. It provides a venue for monographs and multiauthored book-length works that fall within the scope of the journal. The common focus is the emerging field of general evolutionary theory. Such works, either empirical or practical, deal with the evolutionary perspective innate in the change from the contemporary world to its forseeable future.

The examination of contemporary world issues benefits from the systematic exploration of the evolutionary perspective. This especially happens when empirical and practical approaches are combined in the effort.

The World Futures General Evolution Studies and journal are the only internationally published forums dedicated to the general evolution paradigms. The series is the first to publish book-length treatments in this area.

The editor hopes that the readership will expand across disciplines where scholars from new fields will contribute books which propose general evolution theory in novel contexts.

PREFACE

This volume is the product of a joint effort by the members of an international and interdisciplinary research group dedicated to the exploration of what we call the new evolutionary paradigm.

The General Evolution Research Group came into being following an exploratory meeting held in February 1986 on the invitation of Jonas Salk at the Salk Institute in La Jolla, California. The meeting itself was the outcome of preliminary discussions between Jonas Salk and myself during the Discovery Symposium of the Solvay and the Honda foundations in Brussels the previous December. This symposium was held under the scientific aegis of Ilya Prigogine.

The focus of discussion at these two events was the possibility of the mutual enrichment of our understanding of the course of human affairs through the transfer and application of recent theoretical breakthroughs in the natural sciences. The obvious common focus in nature and in society appears to be evolution conceived as irreversible and nonlinear change in domains far from thermodynamical equilibrium. The scientists who met in both Brussels and La Jolla were convinced that evolution in nature and evolution in the human realm have more in common than mere analogies: there are basic isomorphies that point to a consistency at the very heart of empirical reality. The General Evolution Research Group came into being to systematically explore these fundamental commonalities.

This book is the first opus to emerge from the joint work of the research group. A number of further publications are currently in preparation, including the proceedings of the group's meetings in Bologna, Italy. The group publishes individual papers and book reviews in the quarterly journal *World Futures: The Journal of General Evolution* (New York and London: Gordon and Breach Science Publishers), a publication open to all scholars in the general evolution field. The group also circulates research notes, projects, and items of interest in a limited circulation newsletter made available through the Vienna Academy for the Study of the Future.

Ervin Laszlo

FOREWORD

To what extent do human activities stand outside the natural world? Are the conceptions we have of the "laws of nature" applicable to a description of human conduct? These questions have dominated the history of ideas in the Western world; they are at the root of the work of Kant and Hegel, Whitehead and Heidegger, to quote only a few names. Classical science, which started with the "Newtonian paradigm," was centered around the idea of deterministic and time-reversible laws. In this context, the world was seen as a vast automaton; man appeared as being outside nature, as a free agent in a mechanical universe, able to manipulate and exploit his environment.

The classic image of science continues to be propagated with considerable authority. It permeates many aspects of the human sciences, in which rationality is identified with timelessness and equilibrium.

However, at present, our conception of nature is undergoing a radical change toward the multiple, the temporal, and the complex. A new paradigm is taking shape. It is now understood that the behavior of matter under nonequilibrium conditions can be radically different from its behavior at or near equilibrium; and it is precisely this difference that introduces multiple choices, self-organization, and complex dynamics.

There is a close relation between nonlinearity and distance from equilibrium. Close to equilibrium, the description of the temporal evolution can be expressed by linear equations. However, far from equilibrium, we deal with nonlinear equations, which may lead to bifurcations and to the spontaneous appearance and evolution of organized states of matter: "dissipative structures."

It is reported that Bohr stated to Planck that in quantum mechanics, coordinates and momenta cannot be determined simultaneously. Planck answered: "But God knows both!" Bohr's alleged response was that physics dealt only with what *man* could know. The illusion of a com-

plete (infinite) knowledge comes, it would seem, from the historical fact
that classical science started with the study of periodic motions. The
return of the sun and the regularity of celestial phenomena have deeply
influenced man's thought since Paleolithic times. It led to the prototype
of knowledge as expressed in classical physics. But this regularity of
periodic processes is not the general case. The message conveyed by the
second principle of thermodynamics is that we are not living in a world
that can be described in terms of periodic motions. It is an unstable
world, which we know through a "finite window."

In the approach advocated here, rationality is no longer to be identified with certainty, nor probability with ignorance. At all levels, probability plays an essential role. It was, of course, always understood that
exact (that is, infinitely precise) initial conditions correspond to an idealization, but we find it rather unexpected that giving up this idealization
leads to such momentous consequences.

It is not the first time that such a situation has appeared in the history
of physics. The history of science includes both the progressive accumulation of knowledge and the succession of conflicting theories. For
example, it is well known that due to the presence of Planck's constant k, the structure of quantum theory differs drastically from that of
classical theory. In short, we witness here the inadequacy of the ideal
of "complete knowledge," which has haunted western science for
three centuries.

Since the advent of quantum mechanics, we know that probabilistic
concepts play an essential role in physics. What we observe is that they
now play a basic role not only at the microscopic quantum level, but
also at the macroscopic level. We are progressing towards the "open
universe," with its multiple facets, to quote the title of a book by Sir
Karl Popper.

It is the aim of this volume to discuss this recent reconceptualization
of science as well as its possible role in the modelling of human activities. We seek to identify the points which may be of interest in bridging the gap between the "hard" and the "soft" sciences. According to
the classical view, there was a sharp distinction between simple systems, such as studied in physics or chemistry, and complex systems,
studied in biology and the human sciences. Indeed, one could not imagine greater contrast than the one which exists between the simple models
of classical dynamics, or the simple behavior of a gas or a liquid, and
the complex processes we discover in the evolution of life or in the history of human societies. This gap is now being filled. Over the last
decade, we have learned that, in nonequilibrium conditions, simple ma-

terials such as a gas or a liquid, or simple chemical reactions, can acquire complex behavior. This opens the way to new channels for transferring knowledge from physics and mathematics to a variety of other fields.

Irreversible processes have likely played an essential role in inscribing time, so to speak, into matter at the early stage of the universe and producing the information carried by basic biochemical compounds such as DNA. Indeed, a basic feature of biomolecules is that they are carriers of information. This information is quite similar to a text which has to be read in one direction. Biomolecules present, in this sense, a broken symmetry. It is natural to relate this broken symmetry to the broken time symmetry which is expressed in the second law of thermodynamics. Already, in the simple case of the Bénard instability, irreversibility is transformed into pattern. However, the Bénard flow pattern persists only as long as heat is crossing the liquid layer. In contrast, life has an extraordinary degree of persistence, as it originated over 3.4 billion years ago.

We have just remarked on the origins of life, but obviously the most striking of such problems are the origins of the universe. The modern history of cosmology has been a dramatic one. As is well-known, Einstein proposed in 1917 a static model of the universe. General relativity seemed to him to be the ultimate achievement, as it unified gravitation with space-time. But this static image was to be rapidly given up in 1922 as Friedmann showed that Einstein's equations were unstable, and the observations of Hubble pointed in the direction of an evolving, expanding universe. Later, the celebrated residual black-body radiation suggested that beyond this geometrical evolution there is a more basic thermal evolution.

We come, therefore, to the modern 'standard' cosmology. It presents two fundamental aspects: The temperature of the universe increases monotonously when we come close to the 'big bang,' while, according to the adiabatic type of evolution, the entropy of the universe remains constant. We cannot go into details here, but the important point is that the part of the system which is slowly evolving to equilibrium has itself been brought out of equilibrium by a *nonequilibrium process*. Could it be that matter (the baryons and leptons) is the nonequilibrium product of some cosmic nonequilibrium which has also produced the black-body radiation?

There is a basic duality which appears in general relativity; on the one side, space-time; on the other, matter. These two aspects are connected through Einstein's field equations. In recent years, describing the creation of the universe as a "free lunch" has become quite popular. The energy of the vacuum would be transformed into a *positive* energy (matter), and a *negative* energy (gravitational field energy):

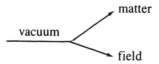

However, if matter is indeed created from the vacuum, this should be an *irreversible, entropy-producing process*. In this view, the entropy of the vacuum would be zero. Matter would be the "contamination" of spacetime, carrying the entropy, generalized by this creation process.

The intriguing new possibility is that the universe may start cold and empty, then heat up through the decomposition of unstable particles, reach a maximum temperature, and then cool down through adiabatic expansion. Similarly, the entropy would start from zero and grow with the creation of matter.

Irreversibility plays a constructive role in the universe. Questions such as the origin of life, the origin of the universe, or the origin of matter can no longer be discussed without recourse to irreversibility. The search for a fundamental time-reversible level in nature seems to have come to an end. Indeed, irreversibility must exist on all levels, or it can exist on none; it cannot emerge in the transition from one level to another.

This leads us to reconsider the relations between man and man as well as the relations between man and nature. Clearly, a social system is by definition a nonlinear one, as interactions between the members of the society may have a catalytic effect. At each moment fluctuations are generated which may be dampened or amplified by society. An excellent example of a huge amplification is the acquisition of knowledge, which in a few decades led from the work of a few pioneers in solid state physics to the information revolution that we are witnessing today. Scientific and technological progress probes the stability of the social system. In this view, there can be no question of the "axiological neutrality" of science. Problems that may arise at the science/society interface can only be solved by understanding the actual complexity of societal processes. If these are not understood, the response of the system may be ultimately a negative one.

Jonas Salk coined the expression "the survival of the wisest." Clearly, the growth of human population requires a new dialogue between man and nature as well as new relations among men. It is the burden of our generation, and the challenge accepted by the authors of this volume, to spell out more precisely the form this new dialogue has to take if we are to overcome feelings of alienation and frustration, which could lead to an emotional rejection of science itself.

Ilya Prigogine

INTRODUCTION

What is general evolution theory? We may properly devote this introduction to an examination of the topic, tracing the theory's intellectual antecedents as well as its present position in the scheme of the sciences.

We shall not make a rigorous distinction between general theories and paradigms. The debates on the nature of scientific paradigms have been many, and they have become increasingly sterile. No useful purpose would be served by choosing one or another of the numerous definitions offered in the literature, or by creating yet another. Since the publication of Thomas Kuhn's book in 1962 on the structure of scientific revolution, the concept has gained wide currency, and despite the lack of precision with which it is often used, it denotes an idea with which most people have adequate familiarity, or at least with which they feel comfortable.

For our purposes it should suffice to note that a general theory is much like a paradigm in that it functions as a conceptual frame for the formulation of specific questions and for particular theories responding to them. A general theory, in this sense, is a "paradigm for" doing science in a given field. A few examples include quantum theory in microphysics, relativity theory in field physics, the modern synthesis in population biology, and Freudian or Jungian analysis in depth-psychology. A general theory can, however, also cross the boundaries of particular fields and disciplines. If it is an interdisciplinary—or, better, transdisciplinary—theory, it can serve as a "paradigm for" investigations in the various disciplines over which it extends. There are not many general theories of this kind, and workers in specialized fields are rightfully skeptical of their validity. However, evolution is one of the very few areas where such skepticism must be suspended until systematic explorations throw more light on the warrant for a general theory centered on its processes.

Prima facie, the study of evolution extends beyond the range of individual scientific disciplines. Scientists implicitly assume—though they do not always explicitly theorize—that out of the probabilistic, quantized ground investigated by microphysics emerged in succession the more measurable and knowable phenomena of the macroscopic world. Cosmologists tell us that the synthesis of matter began at the first 10^{-33} second that marked the end of 'Planck-time' in the life of the universe. Astrophysicists affirm that the synthesis of progressively heavier elements continued ever since, both in stars and in interstellar space. Life scientists assure us that biological evolution took off on this planet sometime between 3.6 and 4.6 billion years ago, and anthropologists claim that hominid species appeared in the last few million years with sapiens emerging as a species capable of language, tool-use, and abstract thought in the span of the last 100 thousand years. Historians and social scientists, in turn, testify that the groups and societies of *Homo,* coded by the symbolic "communicational realities" permitted by the large cranial capacity of our species, began to evolve into today's complex sociocultural and technological societies some 20 thousand years ago. Despite its probabilistic and problematic basis in quantum phenomena, the world we experience managed to work itself up on this planet—and possibly elsewhere in the universe—from common and presumably unified beginnings to its present diversified condition.

That evolution unfolds sequentially does not mean, however, that it is either continuous or internally consistent. Discontinuities between physical-chemical, biological, and human and cultural evolution led to the separation of modern science's various domains of investigation. The differences among the phenomena investigated in the separate domains dominated the awareness of their similarities, at least until recently. Thus we now confront a situation in which there are numerous, highly specialized and independently conducted studies of evolution—studies of the "evolution *of*" particular entities, such as stars, cultures, or insects.

When reviewed in the context of an embracing vision of the evolution of the empirical world, particular studies of the "evolution *of*" things need not prove to be mutually inconsistent. While there is a field specialized in the study of the evolution *of* the cosmos (cosmology), and other fields specialized in the study of the evolution *of* micro- and macro-structures in the cosmos (astrophysics), as well as *of* organisms, species, ecosystems, human beings and social systems on earth (life sciences and social sciences), there could also be a field committed not to the study of this or that unit or entity of evolution, but to the general patterns traced by evolving systems in the cosmos and on earth. The possibility that specialized fields could uncover mutually consistent processes notwithstanding their independent investigations cannot be ruled out. On the contrary, in view of the consistence we observe in nature, it is even likely no matter which of its aspects we may consider. This possibility, then, provides the logical ground for a domain of investigation that we identify as general theory of evolution.

Whether or not the facts, observations, and data accumulated in specialized theories of evolution have the required degree of internal coherence cannot be prejudged. We can neither say that evolutionary processes lend themselves to treatment by general theory, nor that they do not. The question is pragmatic: the proof of the pudding can only be in the eating. If we discover an adequate basis for a consistent general theory without forcing the data into a Procrustean bed of preconceptions, our field of investigation would be as valid as any other more specialized field in the sciences.

But there is more to the project of general evolution theory than the assembly of specialized findings into a consistent general framework. We would not do justice to the nature and role of general theories by assuming that they have merely a passive, metascientific function. If and when specialized theories are connected through a general theory, the nature of the specialized theories themselves changes. This process has been well-described in the literature on paradigms: a new paradigm alters the way in which we look at our data and changes the assumptions that underlie our theoretical premises. General theories of evolution would no doubt exercise such a heuristic function. In virtue of exhibiting the mutual consistency of specialized theories dealing with this or that unit of evolution, they would change our perception of those very entities. They would appear as particular exemplifications of more general classes of events, found also in other theories. The existing theories would then be divested of their comparative isolation.

Thus the exploration of the possibilities for creating general theories of evolution is by no means a trivial pursuit. Its consequences for our understanding of the nature of the world we live in, and of ourselves as part of that world, would be highly significant. The enterprise could, of course, end in failure. But even then, it would change our conception of reality, and our conception of science. For our implicit belief in the consistency of nature—a fundamental tenet of modern science—would be undermined. In confrontation with the compartmentalization of specialized disciplines, faith in this tenet must prove stronger than the territoriality that keeps theories developed within specialized disciplines apart.

When dealing with such emotional factors, it is important that we should side with potentially fruitful hypotheses rather than with narrow-minded preconceptions.

In any event, even the briefest look at the history of the concept of evolution shows that, in this instance at least, the territoriality that is the habitual correlate of scientific specialization is entirely misplaced. The concept does not "belong" to any given field of study; it is the common domain of all empirical sciences. The term itself comes from the Latin *evolvere,* meaning to unfold. It was first applied, erroneously as it turned out, to the development—or "unfolding"— of the full-grown organism from the minute homunculus that was presumed to exist, fully formed, in the male sperm or in the female egg. Later, the concept of evolution became identified with the theory of Darwin in the field of macrobiology. Many scientists, first and foremost biologists, still lean toward the view that evolution is a biological concept, without direct or significant application in other fields. Yet, already in the 19th century, philosophers such as Herbert Spencer and Henri Bergson went beyond biology in attempting to create more general theories of evolution, with application to history as well as to physical nature.

One might, of course, look at such attempts as pure speculation and continue to restrict the concept of evolution to the processes of mutation and natural selection in biological genotypes and phenotypes. We should not ignore, however, that evolutionary conceptions are now emerging in the physical sciences also. With the demonstration that Einstein's cosmological equations have unstable solutions, and with the confirmation of the Doppler effect in astronomy, steady state concepts of the universe had to be surrendered; the states of the cosmos are now conceived as following an irreversible trajectory. From the Big Bang to infinite expansion in the open universe scenario, or to a Big Crunch in the closed universe hypothesis, the cosmos itself is seen to be subject to irreversible change.

Irreversibility, as Ilya Prigogine has justly emphasized in his foreword to this volume, has entered the theories and world conceptions of contemporary physics. Irreversibility is clearly present also at the other extreme of the range of the empirical sciences: in the historical and social disciplines. Aside from theories of cyclical and eternal recurrence, and the positivist disclaimers of the meaningfulness of large-scale patterns, modern social science is firmly based on the concept of historical development. A comparison of contemporary techno-industrial information societies with the still-surviving remnants of hunting-gathering tribes highlights the range, the dimensions, and the statistical irreversibility of the historical process.

Similar concepts have surfaced in psychology and personality theory, where investigations such as those of Piaget have thrown light on the sequential unfolding of perceptual and intellectual capacities in the lifetime of the individual. Most scientists would agree that the human being, no less than cosmos and culture, changes irreversibly over time.

One should not make the mistake, however, of looking at the concept of evolution as a grab bag for concepts of change in the empirical world. It is only on a superficial level that it would seem that there is no process of change that would be meaningfully *excluded* from the concept of evolution. On closer inspection many exclusions become evident. First of all, although evolution is the study of change, it is not the study of all varieties of change. Purely random and entirely time-reversible patterns of change are excluded: evolution concerns exclusively change that is, at least statistically, irreversible. But not even all varieties of irreversible change fall within its compass. To qualify, irreversible change must entail processes that lead to the emergence, or at least the persistence, of ordered structure in space and time. Such processes must constitute an orderly sequence traceable, in principle, from the origins of the physical universe through multiple hierarchical levels and processes to whatever state or process we may wish to study. Evolution, in other words, is the study of progressive, ongoing (but not necessarily continuous and linear) change, leading with at least a statistical irreversibility from the origins of the cosmos to its present state—and (though not necessarily predictably) to its future states.

Evolution cannot be properly restricted to any aspect or phase of this process; it is a universal, even if not an omnipresent, phenomenon. Its range is broader than the range of any of the traditional divisions between scientific disciplines. For such boundary transgressions we cannot reproach nature—it is not obligated to observe the division of labor among scientists. We should, however, do our best to rectify the disciplinary bias that is still prevalent in the investigations of evolutionary phenomena.

There are notable developments within the contemporary sciences dedicated to overcoming disciplinary boundaries and creating transdisciplinary research programs. First and foremost among them is the group of emerging fields known collectively as "systems sciences" or, especially in Europe, as "sciences of complexity." These new sciences deal with the appearance, development, and functioning of complex systems regardless of the domain of investigation to which they belong. They originated with the general system theory pioneered by Ludwig von Bertalanffy, Paul Weiss, Anatol Rapoport, and Kenneth Boulding; and with the science of cybernetics developed by Norbert Wiener, W. Ross Ashby, and Stafford Beer. The pioneer fields were reinforced with the work of Aharon Katchalsky, Ilya Prigogine, and the Brussels school of nonequilibrium thermodynamics; and with the progress of computer-simulated nonlinear dynamics, carried out (following earlier work by René Thom and Christopher Zeeman) by Ralph Abraham, Christopher Shaw, and other mathematical "chaos" theorists. The findings that emerge from the workshops of these scientists prove to have application to a wide range of phenomena; from cosmology, physics, and chemistry to biology and ecology, and from psychology and personality theory to historiography and psychology and the allied social sciences.

In the 1990s, the ground seems to have been prepared for a thorough exploration of the new evolutionary paradigm—for the elaboration of a general theory of evolution. For it now appears that the presupposition of internal consistency among the separately investigated branches of evolutionary disciplines is borne out by the actual results of scientific investigation. Within biology proper, the new concepts of saltatory evolution, advanced in the theory of punctuated equilibria of Stephen J. Gould and Niles Eldredge, disclose a dynamic that is isomorphic in its essentials with the nonlinear dynamic of bifurcations in the complex systems investigated in nonequilibrium thermodynamics and dynamical systems theory. The new cosmologies created by the grand unified theories (GUTs) pioneered by Stephen Weinberg and Abdus Salam, and the new inflationary scenarios developed by Guth, Hawking, Linde, and other cosmologists, also exhibit basically analogous nonlinear dynamics in the discontinuous yet progressive emergence of order and complexity. With pioneering historians and social scientists exploring the application of the nonlinear dynamics of "bifurcating" systems to human society, the major domains of evolution become populated with investigators intent on the creation of new general theories that embrace a wide variety of evolutionary phenomena.

The new wave of interest in the general patterns of evolution is reflected in a spate of high quality literature. In addition to the already "classic" works of Prigogine, Salk, Waddington, and Jantsch, there is Vilmos Csányi's *Evolutionary Systems: A General Theory of Evolution* (Duke University Press, 1989), which sets forth in detail the replicative systems theory that its author sketches in this volume; Eric Chaisson's *The Life Era: Cosmic Selection and Conscious Evolution* (Boston: Atlantic Monthly Press, 1987), a wide ranging work that traces the evolution of matter, energy, and life in its cosmic setting; and Riane Eisler's *The Chalice and the Blade* (San Francisco: Harper & Row, 1987), an innovative study of the evolution of social structures and relations in recorded history. This writer's *Evolution: The Grand Synthesis* (Boston and London: Shambhala New Science Library, 1987) provides a comprehensive review of the basic concepts and principal applications of general evolution theory.

The published literature, and even more the discussions and assessments of research presently under way, suggests that we are approaching a landmark in the history of empirical science. General evolutionary concepts of the world may soon move from their classical position in philosophy and metaphysics to physics and the other empirical sciences. The exploration of the new paradigm, through the creation, criticism, and elaboration of progressively more refined general evolution theories, is a major challenge awaiting the contemporary community of natural, human, and social scientists.

CHAPTER 1

Complex Dynamical Systems Theory: Historical Origins, Contemporary Applications

RALPH ABRAHAM

Editor's Introduction: In this opening chapter Ralph Abraham, a pioneer of the mathematical modeling and simulation of complex systems, traces the origins of the latest mathematical theories applicable to the systems that emerge and evolve in various domains, in nature as well as in society.

A concise and definitive account, the chapter serves as a general introduction to the mathematical underpinnings of general evolution theory, comprehensible to the layman as well as to the specialist. A number of the basic concepts that appear throughout this volume are here described and defined, including the concept of dynamical system itself, and the modeling and simulation of its various stable and unstable states with the help of static, periodic and chaotic attractors and the processes known as bifurcation.

Aware of the practical potential of the mathematical simulation of real-world systems as well as of the mushrooming problems that face contemporary people and societies, Abraham stresses the urgency and importance of researching and developing applications of the mathematical models of complex systems in the human and social sciences.

INTRODUCTION

Since the last glaciation, we have extensive records of some ten thousand years of the struggles of the human species for survival within the ecosystems of Terra. We have coextensive records of the evolution of consciousness, wisdom, intelligence, arts, sciences, and technology. The mutual interactions between these two levels of history have been critical to the survival of our colony up to the present moment, and will

Dedicated to Erich Jantsch (1929–1980)

continue to be critical, as we face the challenges to come. In this essay, we examine the cognitive strategies entwined in the historical records of the sciences, and propose an extrapolation for the near future which may be essential for our survival: *the mathematical acceleration of social theory.*

We will begin with a brief history of the role of mathematics in the development of the sciences since Newton, from the viewpoint of modeling and simulation. Then, we will outline three case studies: dynamics, physiology, and sociology. Finally, we propose an inexpensive project for the accelerated development of a large-scale model of our emerging planetary society, suitable for high-speed simulation by existing supercomputers.

The motivation of this essay, and the proposed project, is the challenge of meeting the oncoming evolutionary crisis, and surmounting it, through a timely increase of our understanding of complex systems and their transformations. For we feel that this increase in understanding will come soon, or never.

HISTORICAL INTERACTIONS BETWEEN MATHEMATICS AND THE SCIENCES

Mathematics is not a science, nor is science mathematical. The applications of mathematics to the sciences involves, in fact, a relatively small part of our mathematical activity, and an even smaller smaller part of our scientific efforts. Yet historically, this interaction has been particularly important in the development of each. This is particularly true since Huyghens, Newton and Leibniz, who were primarily responsible for the cognitive style which dominates scientific theory today.

Applied mathematics, as we may call the interaction between mathematics and the sciences, has two aspects: *modeling and simulation.* Modeling denotes the creative activity of building a mathematical model for a given phenomenon, or experimental domain. It may involve any branch of mathematics in the architecture, construction, testing, and evaluation of a model. Simulation, on the other hand, denotes the operation of an existing mathematical model for purposes of prediction, or study, of the target system. The computer revolution has changed the dominant method of simulation from classical analysis to numerical computation and graphical presentation. We wish now to focus on the modeling aspect of applied mathematics, which was called *mechanics* in ancient Greece.

According to this *mechanical paradigm,* our cognitive strategy in technical matters is mechanical. That is, we understand complex phenomena by constructing models, rather than by verbal, symbolic, or other representations. Models may be physical machines (such as orreries or planetaria), pictorial representations (such as photographs) or mathematical models (symbolically represented, as in $F = ma$). The relationship between the model system and the real target system is a conventional (fictitious) one, and need not be an ideal analogy in order to be cognitively useful. Many different models of the same target system (a *spectrum* of models) may be used at once, to advance understanding. In fact, this may actually *be* understanding. We call this the *mechanistic* approach to science.

This approach differs from that of dogmatic science, in which the model comes, over time, to be identified with the target system. For example, a traditional physicist may assume that the electrostatic potential of Maxwell's model has an actual existence in the phenomenal universe.

Accepting the mechanistic approach, let us review the role of the modeling aspect of applied mathematics in the history of the sciences since Newton.

Throughout the period 1680–1930, there was a growing list of spectacularly good models for physical phenomena. These have become, with surprisingly little evolution since their original creation, the cornerstones of mathematical physics: dynamics of particles and continua, electrodynamics, gravitational theory, thermodynamics, statistical mechanics, quantum theory, and so on. In each case, history follows the same pattern: experimental evidence mounts, cognitive strategies form and dissolve, data are increasingly numerical, models become increasingly mathematical, and so on. Eventually, someone has a revelation or intuitive leap, and theory emerges in a new simplicity of understanding, clothed in a splendid model (Maxwell's equations, Einstein's tensor, etc.) which stands as an ideal model for a long time. In this pattern of punctuated evolution in the sciences, the mathematical models play a key role in the formative stages and cognitive strategies, through interaction with the experimental and theoretical developments. This common pattern is a central point in this essay, and can be learned in detail from a single case study. An ideal case is d'Alembert's wave equation for the vibrating string, which established the dominant modeling style of mathematical physics in 1752.

We will now go on to consider three other cases, one each from the physical, biological, and social sciences.

THEORETICAL DYNAMICS

The word *mechanics* meant model-making to the ancient Greeks, as we have noted above, while the word *dynamics* referred to the medicinal power of plants. In the context of the physical sciences, these two words have become synonymous, and denote the science of force, mass and motion begun by Galileo. From the point of view of mechanics (model-making), the history of mechanics (dynamics of particles and continua) provides outstanding examples of the role of models in the creation of theory. It is very instructive to study them in detail, but here we will be satisfied with a brief listing.

In 1560 or so, Galileo made use of real (physical) models (marbles, inclined boards, leaning tower of Pisa, and so on) to elucidate the basic principles of motion. After creating the calculus in 1665, Newton used it to make mathematical models for the same phenomena in 1685. From the study of these models grew *classical analysis,* one of the main branches of mathematics. The goal of analysis was to obtain predictions (that is, simulated data) from the models (differential equations) by symbolic integration (that is, from explicit functions).

In 1865, James Thompson invented the first mechanical analog computer for the simulation of these same models, providing a second simulation strategy. In the 1920's, Van der Pol began using electronic analog computers for modeling and simulation, and these became fast enough to compete with classical analysis as a practical method. Later, during World War II, they became fast enough to be used as bombsites, simulating trajectories according to Newton's model. Shortly thereafter, digital computers replaced them as the simulation strategy of choice for most dynamic models.

The models created by Newton (coupled systems of nonlinear differential equations) are basic to all the simulations which followed, whether by classical analysis or analog or digital computation.

SIMPLE DYNAMICAL SCHEMES

An outstanding problem of theoretical dynamics is the stability of the solar system. In 1885, Poincaré showed that Newton's methods of classical analysis were inadequate to resolve this fundamental problem. He went on to establish totally new mathematical methods for the study of dynamical systems. These were geometric, rather than analytic, and gave rise to new branches of mathematics such as differential topology. The new methods, applied to systems of ordinary differential equations, are now known as *dynamical systems theory* (or qualitative nonlinear dynamics). They have provided a synthesis of all the outstanding models of the physical sciences into a single modeling strategy.[1]

A *dynamical system* is based upon a *state space*, or geometrical model for the virtual states of the target system. Each point of the state space represents a single, instantaneous state, perhaps through some number of observable parameters. The *dynamic* is a infinitesimal rule of evolution: each state is characterized by a unique evolutionary tendency, described by a velocity vector.

The behavior of these mathematical systems is well-known, through three centuries of experimental and theoretical findings. A given initial state evolves along a unique trajectory. After a temporary phase, the *transient response*, this trajectory approaches asymptotically to a limit set called an *attractor*, and a dynamical equilibrium is attained. These occur in three flavors: static, periodic, and chaotic.

Static attractors, also called *rest points*, have been extensively applied since the time of Newton. A system under the influence of a static attractor approaches the final destination and slows to a halt.

Periodic attractors, also called *oscillations*, have dominated dynamics for the past century. A system approaching an oscillation will behave more and more like a perfect oscillation as time goes on.

Chaotic attractors, also called *strange attractors*, are newly discovered, and provide for an understanding of many kinds of aperiodic behavior. Much is now known to be signal, which was previously considered to be noise.

In a given dynamical system, there are usually several attractors. As each initial state will evolve to one of them, the state space may be decomposed into sets sharing the same final fate, which are called *basins*. The basins are divided by *separatrices*. The state space, with the attractors, basins, and separatrices drawn upon it, is called the *portrait* of the dynamical system. This portrait comprises the full understanding of the dynamical behavior of the model, at least as far as long-run prediction is concerned.

Most useful models contain adjustable constants, or *control parameters*, which may be used to adjust the dynamic on the fixed state space. Such a model is called a *dynamical scheme*. As the controls change, so does the portrait. The *response diagram* of the scheme is a graph showing the dependence of the portrait upon the control parameters. The response diagram is the master map which gives this kind of model great power in applications. Points in this diagram where the portrait changes in a particularly significant way are called *bifurcations*.

Catastrophe theory has provided excellent pedagogic examples of response diagrams for various schemes, establishing their importance as graphic representations in many scientific disciplines. Further, it demonstrates the usefulness of mathematical theory in these applications, as the theory *excludes* many bifurcations which might otherwise be expected.[2, 3]

For our present purposes it suffices to observe that dynamical schemes unify all the best-known models of the physical sciences within a single modeling strategy.

THEORETICAL PHYSIOLOGY

After the introduction of dynamical systems theory by Poincaré in 1882, and the maturation of mathematical physics from mechanics to quantum theory, a disastrous gap opened between pure mathematics and the sciences. Although mathematical physics was two centuries old, biological science had hardly begun. Thus, unlike physics, biology was forced to evolve with little support from mathematical models. Of course a few biologists had extensive mathematical training, perhaps from backgrounds in physics or engineering. But the mathematics which had evolved in that arena was not ideally suited to biological modeling. So overall, one might say that theoretical biology and mathematics were both retarded by 50 years or so by the lack of interaction. Only in the past twenty years has there been a significant interaction between theory and modeling, and by now the journals of mathematical biology are filled with very sophisticated models.

The bulk of these models are, in fact, simple dynamical schemes. And their style is much influenced by the historical models of the physical sciences. And yet, when we try to transcend the reductionist models for isolated parts of whole systems, the strategies of the physical sciences fail us. Physical systems are too simple to guide us. Thus, new mathematical strategies have recently evolved for modeling the complex systems encountered in biology, such as general systems theory, systems dynamics, nonlinear control theory, urban dynamics, cybernetics, and so on.

One such strategy, a straightforward extension of the dynamical systems theory of Poincaré to complex systems such as networks and membranes of simpler systems, is called *complex dynamical systems theory.* It evolved in efforts to create high-fidelity models of physiological systems, such the endocrine control systems for the regulation of sleep, eating, stress, and immune responses of mammals.[4]

COMPLEX DYNAMICAL SCHEMES

Given two dynamical schemes, each with its own control and state spaces, a simple kind of *coupling* may be defined by a function from the state space of the first to the control space of the second. The first is the *driver* in this coupled system, as its states operate the controls of the other, *follower* system. By making such couplings among several simple schemes, coupled networks may be built. These provide an excellent modeling strategy in scientific areas where reductionist experiments have led to good dynamical models for the individual parts of a whole system. Each part-model is characterized by a response diagram, which may be well-mapped through extensive computer simulations.

Then the challenge to the theory of complex dynamical systems is *to predict the behavior of the complex system, from a knowledge of the behavior of the parts, and their couplings.*

At present, this theory is in its infancy. Even if the component schemes are stable linear ones, as is frequently the case in systems dynamics for example, the behavior of the complex system may be chaotic. Yet the emerging theory of bifurcations of dynamical schemes is very promising here, as it provides the beginnings of an encyclopedia of atomic bifurcations, of which all response diagrams are made. Viewed as exclusion rules, this encyclopedia may be very helpful in interpreting the results of computer graphic simulations of large-scale complex models. As it grows, a useful theory will become available, and although the behavior of a complex may not be predictable from the behavior of its parts, *it may be obtainable from an affordable amount of computer simulation.*

Other contributions to complex dynamical systems theory may be expected from differential topology and geometry, and practical experience will accelerate when personal supercomputers become available in the near future.

THEORETICAL SOCIOLOGY

This subject, beginning its ascent half a century behind that of theoretical biology, may be expected to grow at a faster rate. For the gap between mathematics and the sciences has been bridged here and there. Thus, the further advance of social theory could be meteoric, if it makes uninhibited and interactive use of innovative mathematical models in the spirit of Newton, Euler, the Bernoulli's, d'Alembert, and so on. For 70 years sufficed for the creation of mathematical physics as we know it today, while a like period in the history of mathematical biology advanced us relatively little.

Due to the explosive growth of social problems, *the fastest possible advance of social theory,* including an adequately predictive model, is mandatory. Thus, we need to nurture the maximum interaction between the ingredient subjects (complex dynamical modeling and simulation, pure mathematics, all of the social sciences, computer science) with adequate resources.

In the growth of social theory, what sort of mathematical models might be useful? Just as the simple dynamical schemes of physics had to be extended to the complex schemes of physiology, further extension may be necessary to build successful models for a planetary society. One modification has already been introduced by Stephen Smale, in his microeconomic model for a trading society.[5] In this model, the *dynamic* (that is, the rule of evolution) is specified not by a unique velocity vector at each point in the state space, but by a cone of favored directions instead. Another extension which may be necessary for the modeling of very large and complex systems is a hierarchy, or *spectrum,* of models.

SPECTRAL DYNAMICAL SYSTEMS

By this invented phrase we mean a whole family of complex dynamical models for the same target system. For example, we may have a hierarchy of models, ranked by differences of physical scale in the state space: microscopic, fine-grain, coarse-grain, macroscopic, thermodynamic, and so on. Or, we may have parallel models of the whole system, but seen from the perspective of disjoint local regions. There may be similar models, distinguished by separate hypotheses, decision strategies, policy-making styles, and so on.

This situation is already familiar, not in modeling practices, but in the verbal analyses of social systems. These are parsed by aspects belonging to separate subjects, such as political analysis, economic description, resources, needs, climate, foreign interactions, and so on.

Thus, we may foresee a further extrapolation of dynamical model structure, in which the cognitive styles already firmly fixed in the various social sciences may separately be embedded, each within one of a spectrum of interlocking complex dynamical models. The models of the spectrum must be made in a universal strategy and style so that they may be successfully combined, or coordinated, for purposes of computer programming, for simulation and prediction, for policy-making, and so on.

The master map or *hypermodel* which coordinates this spectrum of models will probably be a known structure from differential topology or geometry. But all this can be elaborated only in the context of the actual future of mathematical social theory.

PROSPECTS FOR AN ACCELERATED DEVELOPMENT OF A SUCCESSFUL PSYCHOHISTORICAL MODEL

We believe that a successful model of planetary society is an attainable goal for our species, complete with accurate predictions for millenia, simple models for chaotic states and transformations, and short lists of alternative futures at the bifurcation points of psychohistory. Indeed, the achievement of a satisfactory social theory, following in the footsteps of physics and biology, *must* provide us with such a model, and perhaps the extension of natural intelligence by the computer revolution is a necessary prerequisite.

However, we may not wish to wait for a century or two for the spontaneous development of this model, from science fiction to the boardroom computer. Indeed, we may not be able to. So we must ask: what are the prospects for the intentional acceleration of this natural development by a large factor, such as ten?

Certainly the exigencies of World War II created maximum acceleration efforts for various physical technologies, such as radar, rocket propulsion, and nuclear reactions. The respective strategies of England, Germany, and the United States for these accelerations were very similar: draft the best people, combine them in an isolated think tank with extensive resources and all the funding that can be spared, provide inspiring leadership and desperate motivation, and hope for the best. In all three cases, luck prevailed.

The history of analog and digital computing machines provides a second precedent. Several centers in England and the United States gambled on different strategies and took their chances, as in roulette. Here too, luck prevailed.

Our current situation may be very similar to wartime, but with all of us on the same side. As the battle for survival intensifies, the defense budgets of the world may be redirected to a desperate program to accelerate the development of the psychohistorical model. The earlier wartime efforts may serve as the organizing plan for a new crash program.

Yet those were based upon applications of sciences with theories already well developed. It may not be possible to accelerate the early stages of emergence of theory. At least, we cannot guess how long this might take.

We have much at stake. *Should we trust to luck?*

CONCLUSION

The history of the three sciences from the point of view of mechanics (model-making) has been considered, to suggest the possible importance of complex dynamical models and supercomputer simulations in the development of social theory, and an adequate model for supporting the emergence of a peaceful planetary society.

It appears that mathematics, computer science and the social sciences are poised for a rapid growth. But the normal rate of this growth may be much too slow to assist us in coming crises. We have no precedent for the intentional acceleration of the formative stages of a science.

As Einstein said: *One can organize to apply a discovery already made, but not to make one.*

It is time to begin.

BIBLIOGRAPHY

1. Abraham, Ralph H. and Christopher D. Shaw, *Dynamics, the Geometry of Behavior*, Aerial Press, Santa Cruz, CA, 1982.
2. Thom, Rene, *Structural Stability and Morphogenesis*, Benjamin, Reading, 1972.
3. Zeeman, E. Christopher, *Catastrophe Theory*, Addison-Wesley, Reading, 1977.
4. Abraham, Ralph H. *Complex Dynamical Systems*, Aerial, Santa Cruz, in press.
5. Smale, Steve, Dynamics in general equilibrium theory, *Amer. Econ. Review*, 66 (1976), pp. 288–294.

CHAPTER 2

Chaos and Transformation: Implications of Nonequilibrium Theory for Social Science and Society

DAVID LOYE

Editor's Introduction: The discovery that order arises out of chaos is as old as systematic thinking about the nature of reality: it was present already in the sixth century B.C., in the speculations of the Ionian natural philosophers about the origins of the universe. The basic concept is now rediscovered in the contemporary sciences, especially in cosmology, nonequilibrium thermodynamics and dynamical systems theory. (It is noteworthy that Prigogine's and Stenger's classic overview appeared in English under the title "Order Out of Chaos".) But, rather than assuming that chaos reigned only in the beginning of the cosmos—where, according to current cosmologies of the Big Bang with their inflationary scenarios and grand unified theories it did reign indeed—the contemporary sciences discover that chaos gives birth to order over and over again, as complex systems become unstable and manifest the discontinuous and nonlinear phase changes known as "bifurcations."

In his broad-ranging overview, Loye outlines the significance and the many applications of the new theories concerned with the emergence of order out of chaos. From chaos in the computer-simulated models of mathematicians, to chaos in natural systems to chaos in human society, the new theories apply to a wide variety of phenomena. There is, to be sure, not one single monolithic "chaos theory" but many theories of chaotic—more exactly, near-chaotic—states. (In entirely chaotic states in one or two-dimensional systems the "universality equations" discovered by Feigelbaum apply—memory of the original global structure of the system is lost.) In states near chaos, and in transitions to and from chaos, a number of remarkable phenomena occur as systems become turbulent and sensitive to minute changes in internal and external parameters. In such states, governed by so-called "chaotic attractors", transitions take place to new forms of order. This process, as Abraham has pointed out in Chapter 1, is rich in implications for our understanding of both natural and societal phenomena.

Loye illustrates and underscores the applications of the new developments in mathematical theories of chaos for the social sciences with an assessment of the practical values of this endeavor, and with a review of work already in progress.

11

A rigorous new scientific perspective is beginning to emerge of how order gives way to "chaos," order is discovered within "chaos," and order is again created out of "chaos." The appeal of this view is both theoretical and practical. For nonequilibrium or "chaos" theory not only generates the kind of excitement that generally foreshadows major scientific advancement.[7] It may also, at a potentially chaotic juncture in human evolution, offer us a much clearer understanding of what happens, can happen, and can be made to happen in a time of mounting social, political, economic, and environmental crises.[66]

While one may intuit such a connection, to go from the liquids and gases of nonequilibrium theory in natural science to the newspaper headline reality of *the world problematique* requires a transciplinary leap that is impossible to make without certain "stepping stones." The purpose of this paper is to provide a bridge from natural science to social action by focusing primarily on the intervening requirements of a new social science. Specifically, we will examine: 1) the realities of social breakdown and potential chaos in this late 20th century world, which present the need for new action-oriented social theory; 2) the "breakthrough" nature of "chaos" theory in natural science; 3) the originating vision of social science as a tool for social problem-solving; 4) the gap between social science's formative hopes and contemporary performance; 5) the roots in modern social science of a social equivalent to natural scientific "chaos" theory; and 6) examples of current social "chaos"/"transformational" theoretical works and works in progress.

"CHAOS" IN CONTEMPORARY SOCIETY

The systems breakdowns in the contemporary world to which disequilibrium or "chaos" theory potentially relates are of two types. One is the microcosmic social reality of crises and discontinuities affecting increasingly larger segments of the world population. Among these potentially chaotic developments are financial crises already affecting Latin America, Africa, and much of Asia; food crises ravaging regions of Africa and South Asia; worldwide political and military crises, exacerbated by the threat of proliferating nuclear weaponry. Underlying and driving these discontinuities and crises is the impact of changing technologies, of the desertification of productive land and general ecological devastation, of the growing gap between rich and poor economies, the ever-larger diversion of resources needed for human services into armaments, the pressure of the world population explosion. Such crises and pressures drive the breakdown of systems that can lead to states of social chaos.[9,32,60]

The other type of problem is macrocosmic: the great, overriding churning of history and acceleration of evolutionary forces that has led to the characterization of ours as the "age of disintegration"[47] or the "age of discontinuity"[13]—but also as a "crucial epoch"[32] during which the future rides upon our decisions and our will. Viewed from this wider perspective, chaos is not solely negative. It is not just the destruction of what exists—and here the threat of the ultimate chaos of nuclear holocaust, ending human evolution, is the prime example of our fears. Chaos is also potentially positive. That is, the hope for our future is that out of the crises of our time may emerge what has been called "the great transformation,"[53,24] or a new, more positive direction for human evolution.

How can "chaos" theory help in this central task? What it offers at this critical evolutionary juncture is the first *transdisciplinary* understanding of bifurcational and transformational change. But to achieve this new understanding, social scientists must understand natural scientific "chaos" theory, natural scientists must understand the social scientific potential, and both must better understand how advancement at both levels relates to the overriding evolutionary challenge.

"CHAOS" THEORY IN NATURAL SCIENCE

As most natural scientists are aware, there is actually not one, but rather many kinds of disequilibrium theory, for which "chaos" is only one of several names. In the main, two streams are evident. One form is a precise, reasonably well-developed aspect of the mathematical study of dynamics.[1] With its comprehension limited to specialists, this theory acts as a main tributary into the other form, which, like a river swelling in floodtime, is today being advanced throughout natural science by generalists as well as specialists. This broader kind of theory—also called "nonlinear dynamics," "bifurcation" or "self-organizing" theory, but to which the term "chaos" theory is applied by those seeking a unifying image—is still at an early stage of development. But already discernible out of this still very loose cross-disciplinary clustering of concepts, languages and personalities from many fields of the physical and biological sciences is the emergence of important commonalities.

A useful entry point into the world of "chaos" or disequilibrium theory is Cornell physicist Mitchell Feigenbaum's discovery around 1976 of two curious things about some simple mathematical equations. Working with a hand calculator and then a desk-top microcomputer Feigenbaum found that if he kept feeding the results of his calculations back into the computer, over and over again, for a time there emerged the predictable results one would expect. Then came the first surprise. Beyond a certain point of repetition, the results began to change in what at first seemed to be completely unpredictable (or chaotic) ways. But then Feigenbaum noticed there were actually patterns appearing within this jumble of numbers. In other words, he found what appeared to be a kind of predictability, in the form of recurrent patterned order appearing out of chaos.[21]

What Feigenbaum and associates discovered was only a fresh set of clues in the investigation of what increasingly looms as one of the central scientific mysteries of modern times. In contrast to the relatively static nature of society before the Renaissance, the striking feature of the centuries since then has been change—technological, social, economic, political, scientific. To understand this upheaval on the basic level of material reality, Isaac Newton began the study of the *essence* of change in Western scientific terms. This became the field of *dynamics,* which over the years slowly grew as a rarified, inner circle interest mainly of mathematicians and physicists. Slowly there accumulated a set of mathematical concepts, expressed in terms of an increasingly precise special language. Change was carefully tracked, for example, in terms of ''static attractors'' and ''periodic attractors'' within a ''state space'' alive with ''trajectories'' and the tensions of a ''vectorfield.'' In this way, for the first time, change was defined in ways that made possible its measurement, its modeling, and—of the most practical importance—prediction of its directions, effects, and rate of speed.[1]

Then came the eruption of an unsettlement foreseen originally by the great mathematician and dynamicist Henri Poincaré in the last century. In 1962 Edward Lorenz, in an early computer graphics modeling of natural phenomena at MIT, identified a third type of tensional nucleus, the ''chaotic attractor.'' Working with the practical problem of how to improve weather forecasting, Lorenz had succeeded in programming the computer to simulate air movement above the earth. What appeared on his screen was both fascinating and greatly disturbing. For as he observed this first simulation of the operation of a chaotic attractor he saw what appeared to be a complete breakdown of predictability. Not only did the scientific assumption since the time of Newton of an increasing control of ourselves and our world through increased predictability seem shattered, but also any hope of improvement for weather prediction—or for that matter, any other kind of forecasting.[35]

By now "chaos" was, so to speak, in the air. The earlier spread of the dynamicist language and point of view throughout much of natural science had generated the beginnings of many other investigations of "chaos." Numerous physicists and chemists, including Onsager, DeGroot, Katchalsky, and most notably Prigogine had, in exploring non-equilibrium thermodynamics, identified a widening net of processes for "chaotic" states.[28] This state they fitted within a thermodynamic view of systems dynamics involving entropy and negentropy, or the running down and building up of systems. They made use of the idea of bifurcation, or the branching of phenomena into other forms during chaotic states. In chemistry, in pursuit of an understanding of pattern formation in compounds, Ilya Prigogine advanced the remarkably productive idea of "dissipative structures," or the formation of order out of chaos through "autocatalysis" and nonlinear interactions in the form of feedback loops with components influencing one another through "cross-catalysis." Important in Prigogine's thought was also the concept of "nucleation" to express a key aspect of the idea of "attractors."[54]

Moving higher in the order of complexity, crossing the boundary between inanimate matter and life, in biology the same kind of pursuit has been underway. There is, for example, the theory of biologic macroevolution advanced by Eldredge and Gould. An alternative to the standard interpretation of Darwin, this theory posits unstable or chaotic states of "punctuated equilibrium," during which "peripheral isolates" may eventually transform a system into another form.[19] There is the development by Vilmos Csányi of a comprehensive new general theory of evolution incorporating disequilibrium theoretical concepts.[10] This work includes, in Chapter 5 herein, the articulation by Csányi and associate György Kampis of "autogenesis" as a key conceptual expansion of Prigogine's "autocatalysis" and Maturana and Varela's "autopoiesis."[11]

Hovering over these, and all other fields, is general systems theory, which has attempted to detect commonalities of structure and dynamics across fields from its early statement by von Bertalanffy,[68] and later development by Boulding,[6] Miller,[46] Laszlo[28] and others. Here the investigation centers on the key concept of the "self-organizing" capacity of all open systems—or "living systems" in Miller's terms. While a system "appears as irregular or chaotic on the macrocosmic scale, it is, on the contrary, highly organized on the microcosmic scale," Prigogine notes.[54,p.141] In other words, self-organizing is the capacity of open and living systems, such as we live in and we ourselves are, to generate their own new forms from inner guidelines rather than the imposition of form from outside.

It is this concept that also aligns the interests of disequilibrium theorists with evolutionary theorists, as new works by Csanyi[12] and Laszlo[33] are articulating. A critical aspect of both evolutionary and self-organizing theory is the difference between how self-organizing operates "out there" and within our own minds, or what happens during and after the cross-over point between the natural and the human world. This focus on human cognition, which shapes human action, is being explored by Maturana and Varela.[42] Further closely allied to, and of great influence at this theory-building juncture, is the field of cybernetics, opened by the work of Wiener,[69] providing the concepts of feedback loops, now used in all these fields, and the ubiquitous ideas of a systems stabilizing negative feedback and a destabilizing positive feedback or feedforward.

THE POTENTIAL OF SOCIAL SCIENCE, AND THE GAP BETWEEN ORIGINATING VISIONS AND CONTEMPORARY PERFORMANCE

As similarities between the "chaos" being examined on the natural scientific level and potential chaos on the social problem level have become apparent, enthusiasm for applying the new theory to the global societal challenge has arisen. However, as we have indicated, between the development of disequilibrium theory in natural science and the problems of late 20th century human society lies a great open territory of much complexity to be traversed. This is the multifaceted realm of social science, which can only be crossed through fashioning a social theoretical equivalent to natural scientific "chaos" theory. Both why this must be done and how it can be done is indicated by the historical three stage framework for the development of science.[54,33,1]

Our earliest or first stage scientific theories—exemplified by the physics of Aristotle and early modern thermodynamics—focused primarily on so-called steady or equilibrium states. A discernible second stage of scientific theory begins with the recognition of periodic fluctuation—that is, the operation of oscillations whereby systems move in and out of (but still remain near to) the equilibrium state. The third stage—to which science has only begun to address its attention within very recent times—is this exploration of states of extreme instability, so-called chaos, where true rather than only quasi- or illusory systems transformation may occur.

It is this aspect of "chaos" theory as the third stage—i.e., the innovative thrust, the leading edge—for natural scientific theory that bears importantly on the question of its relevance to the problems of social crisis. For traditionally what natural science opens up becomes the territory of social science (e.g., the progressively more rigorous quantifying of social science following the lead of mathematics and physics). It is the recognition of this progression that excites both natural and social scientists with the social scientific potentials of "chaos" theory.

Ralph Abraham, as shown in Chapter 1, for example, thinks a modeling strategy known as modular dynamics may "provide civilization the means to transcend coming crises."[2,p.2] Ervin Laszlo thinks "humanity could gain a significant degree of control over the evolution of its societies precisely at the time such control is needed."[31]

Such statements reflect more than recognition of the "third stage" scientific potential, however. For it is at this point that the fundamental difference between systems theory on the natural and on the social levels becomes pressing. This difference is the *normative* aspect of social theory, or the requirement of attention to the systems quidance question of *ideal* developmental forms that must be the prime concern of all policy makers.[41,48]

On the natural science level, "chaos" investigators may—as they do—study the patterns within proliferating masses of numbers or wildly gyrating chemical processes with relatively little compulsion to rate these patterns as "good" or "bad," or to consider whether they eventually lead to "good" or "bad" stable states. But when one crosses the boundary into the arena of life forms, the case becomes radically different. Now it is our own lives we are investigating, raising inescapable questions such as whether these lives are threatened or protected, fulfilled or unfulfilled—and whether what we do with our lives drives our species toward distinction or extinction.

With both the third stage potential and the normative requirements for social science in mind, it is now possible to project unusual benefits for an investment in the development of a social equivalent for "chaos" theory. For example:

1. *Benefits of improved forecasting.* While natural scientific "chaos" study has uncovered specific limits for predictability during chaotic or maximally transitional states, it is also discovering new possibilities for improving forecasting within these limits by identifying patterns that foreshadow either impending chaos or potential order out of chaos. This capacity suggests how world problems may be alleviated by using such theory to develop far more effective "early warning

systems" for identifying impending food, financial, political, and environmental crises. Even more important, it indicates the potential, now so generally lacking, for identifying productive routes *out* of such crises.

2. *Benefits of improved interventional guides.* One of the greatest problems faced by responsible leadership is in knowing precisely where, when, and how to intervene to either prevent problems of social, economic, political, or ecological "chaos" from arising, or once at hand, to alleviate or solve them. The "chaos" theoretical capacity for expressing movement in mathematical terms that can be projected in computer graphics makes it possible to reduce vast quantities of otherwise confusing data into a comprehensible form. This could both simplify and dramatize communication of problems and solutions leading to more swift and effective interventions.

3. *Benefits of participatory rather than authoritarian problem solutions.* Rather than mobilize the intelligence and energies of the mass of non-experts and non-elites to intervene on their own behalf, the traditional tendency during times of trouble has been to turn to the easier and seemingly quicker route of isolated expert, elite, or authoritarian solutions. Such solutions, however, can exacerbate or bottle up the problems, while creating a new range of difficulties. Again, it is possible to visualize that the chaos theoretical approach, through a new capacity for bringing hitherto incomprehensible complexities within human comprehension, could provide a new way of bridging the gap between expert and non-expert making possible the effective joint solution.

4. *Benefits of providing a clearer sense of system goal states or prohuman images of the future.* A problem remarked by many perceptive futurists is how, in contrast to the fervent visions of a better future that animated the revolutions and reforms of the 18th and 19th centuries, in this century a confused and fearful humanity seems to be running out of vision.[52] As a consequence, most efforts to attain a better future now seem to flounder or misfire. The significance of "chaos" theory is that for the first time in the evolution of our species we are beginning to develop a way of scientifically grasping not only the oscillatory return to systems equilibrium or quasi-equilibrium, but also how to break out of our social and evolutionary stalemate into a world of incredible possibilities. As with all other forms of liberation, this new liberation and indeed creation of a New Mind must surely revitalize our image of the future and positively recharge motivating goal states.

To attain such third stage/normative benefits, however, social science must move much more rapidly in two critical regards. It must, first, catch up with the developmental lead of natural science. Part of the problem here is that social science is still largely immersed in the first and second stages of scientific development. As made evident by the difference between the projections of its pioneers and the actualities of social scientific development, only with the greatest difficulty, and with a significant time lag behind natural science, has social science begun to move with even minimal certainty into third stage science.

The other part of the problem also derives from the historical dependence of social science on natural science. It is that social science has doggedly persisted in trying to apply non-normative first, second and even third stage models, which prevail in natural science, to the normative world of the human being, human society, and the embracing whole of our ecology. As numerous critics have noted, this kind of reductionism leads to the science-without-values impasse of Auschwitz and Hiroshima.[9,61]

The practical effect of this double lag has been the ineffectiveness of social science in dealing with mounting problems of discontinuous change in the real world. Lacking widely acceptable modernized *and* intelligible theories of social change (the third stage requirement) and of *directed* social change (the normative requirement), social scientists have largely been dealing with the problems in heuristic, ad hoc, and all too often ineffectual mini-system and microtheoretical ways.

So what an evolutionary-process-oriented "chaos" theory now offers is the promise of a major step toward the formulation of the kind of third stage and normative framework that, as we will see, was often visualized by the great formative social theoreticians, but then fell by the wayside as social science bogged down in scientific developmental stages one and two. Now, if social science is to move on, certain problems must be recognized and overcome.

One seemingly minor, but in fact quite important, problem bearing on the articulation of both third stage and normative potential is linquistic. This is the term "chaos" itself. Its use in the social context may seem to some an exaggeration for dramatic or rhetorical rather than scientific purposes. For what most of us see when we consider the social world about us is not chaos, but a mixture of order and disorder. The same kind of difficulty exists with the use of "nonlinear dynamics" as an alternative. For again what is evident in the social context is the operation of linear as well as nonlinear dynamics, with the overriding interest being in dynamics per se.

For such reasons, we suggest "transformation theory" as a general designation for a social equivalent to natural scientific "chaos" theory. As exemplified by a theory of this new genre briefly described later, Eisler's "cultural transformation theory,"[18] such a term subsumes both chaos and order and focuses on the central aspect of social theory to which those who comprehend the chaos theoretical potential are deeply responsive. This is the idea of *transformation* as a process out of or through which order gives way to, chaos, and chaos again leads to order. The term "transform" is also meaningful on both hard and soft science levels, and because of appealing, non-threatening connotations would offer great communications advantages.

Much larger, however, is this great underlying problem that neither social scientists nor policy makers have yet satisfactorily solved: how despite formative visions of social theory that might provide guidance for social policy, social science still generally remains isolated from the suffering body of society itself.

One reason for this isolation again, curiously, has a linguistic aspect. It is that over the past 150 years of social theory development more and more terms for the same phenomena have emerged. Added to this have been the imprecisions of more and more languages in translation, with conceptual qualifications piled upon qualifications, until "grand"—or whole systems sensitive—theory development became such a strain that it largely dead-ended in the prolix attempt of Talcott Parsons at a synthesis of early, more vivid work.[50] The net effect has been that, while the problems it was supposed to help have exponentially increased, social theory has tended to become either the plaything of an increasingly lonely academic elite or the weapon of comparably lonely activists with little knowledge of or experience with science.

It is in regard to this persisting developmental barrier that the "chaos" theoretical potential for social scientific advance presents its most startling and engrossing aspect—and this realization can burst upon one with considerable force. For if, guided by isomorphisms across levels, one can avoid the pitfalls of the past, it may at last indeed be possible to get through to—and act upon—the problems! In keeping with the general systems theory approach, for example, of Miller in *Living Systems*,[46] the basic strategy would be to compare change phenomena on the social level with change concepts from the new natural scientific investigation of "chaos," looking for matches and mismatches, and then field testing and re-analyzing findings. If through such an approach one can bypass the problems of fragmentation, isolation, and fruitless quarreling for theory developed in the old way for social science, it is possible that the practical goals of the pioneers of social science may at last be within humanity's grasp.

But to state the possibility in this way not only touches one of social science's great historical sensitivities but also, with many, invites immediate dismissal of the whole idea. For isn't this the old ogre of positivist reductionism on an incredibly sweeping scale? Won't this idea of boldly and directly using the concepts of natural science to guide social theoretical development only further entrench us in the old problem of non-normative models? Won't we be inviting back problems like those of Social Darwinism, or more recently "computerism," where concepts that work or are true for a "lower" level are used to falsify matters at a "higher" level?

THE ROOTS OF SOCIAL "CHAOS" THEORY

In fact, "chaos" theory is not a new or alien notion to social science. Its roots in social science go back at least to early articulations of dialectical theory in the first proto-scientific work on forecasting, the ancient Chinese Book of Changes,[26] and to the early Greek philosophers, notably Heraclitus.[62] In modern times, "chaos" or "transformational" theoretical questions lie at the core of Hegel's philosophy of history as well as the multi-faceted dialectical theory of Marx and Engels that initiated the formative dialogue for the development of modern sociology.[25] Moreover, Prigogine himself cites Auguste Comte, Emile Durkheim and Herbert Spencer as the forerunners of his concept of dissipative structures, referring particularly to Durkheim's concept of moral density as a precondition of the division of labor.[5]

In other words, rather than reductionism we confront a case of cross-fertilization, with the mutuality of benefits for *all* levels of science that this implies. For closer scrutiny of Marx, Engels, Weber, Pareto and others reveals they were all grappling with isomorphically the same questions of change as modern "chaos" investigators in natural science. For example, one may detect isomorphisms worth investigating between, on the one hand, the cross-catalysis observed by Prigogine in chemistry, and, on the other, concepts of Durkheim,[15] Sherif,[59] and others of solidarity, conformity, norm formation, and other ideas bearing on the "nucleation" of group formation. Prigogine's observation of autocatalysis as well as Lorenz's and Abraham's computerized projections of the operation of the chaotic attractor seem suggestively linked to what Max Weber was trying to define with his concept of charisma.[69] Durkheim's famous concept of anomie is an attempt to describe the psychological effects of the breakdown of norms and social expectations that characterize social chaos states.[66] A case can also be made for interpreting alienation as a comparable effect of reaction to the restraint of too much order.[36]

The equilibrium-disequilibrium alternation, basic to this kind of theory, has seldom been stated better than as follows by pioneering sociologist and economist Vilfredo Pareto: "the governing elite is always in a state of slow and continuous transformation. It flows on like a river, never being today what it was yesterday. From time to time sudden and violent disturbances occur. There is a flood—the river overflows its banks. Afterwards, the new governing elite again resumes its slow transformation. The flood has subsided, the river is again flowing normally in its wonted bed."[49,p.555]

In psychology, the action research and theory of Kurt Lewin—whose influence was pivotal in the development of social psychology and its application to the problems of prejudice, race relations, group dynamics, worker productivity and morale, leadership development and leadership styles—fairly explodes with dynamicist concept equivalencies. And again in Lewin we find an articulation of the social aspects of "chaos" theory in his famous perception of directed social change as a three-step process. At first there must be the "unfreezing" of the prevailing state, which requires action to "break open the shell of complacency and self-righteousness." This is followed by second stage "moving to new levels." But then there must be a third stage "re-freezing" at this new level to prevent a regression to the former or original state.[34,pp.228,229]

One encounters such thinking in the disequilibria-oriented work of Pitirim Sorokin in sociology[63] and the formation of dialectical psychology by such figures as Klaus Riegel in the United States and S. L. Rubenstein in the Soviet Union.[55] It is also evident that "chaos" or "transformational" thought was a concern of evolutionary-oriented historians such as Oswald Spengler[64] and Arnold Toynbee[67] as well as Sorokin, whose theories of how cultures are born, grow, meet or do not meet challenges, decline and even die, were, on the broadest possible scale, attempts to grapple with problems that now confront developers of a social equivalent to "chaos" theory.

The fact that these connections have been so quickly forgotten, or are only sporadically pursued or emphasized, is again evidence of the hold of stage one, stage two and anti-normative paradigms on social science. But now the development of theory adequate to the challenge of the times and the formative social thinkers' visions may be at hand.

CHAOS/TRANSFORMATION THEORETICAL
WORK IN PROGRESS

Already possibilities for such a body of theory can be glimpsed in a variety of current works and works in progress. From the perspective of translating "chaos" into "transformation" theory, an important beginning was René Thom's catastrophe theory, which has been applied to forecasting and the analysis of a variety of societal tasks.[65] Zeeman and others, for example, have shown how otherwise inexplicable dynamics in the case of stock market activity, prison riots, "doves" versus "hawks," political opinion change, and even brain functioning operate according to the mathematical predictability of "catastrophic" shifts during disequilibrium states.[71] Likewise, Prigogine's dissipative structures theory has been applied by Allen to the analysis of urban and transportation systems growth patterns.[3] And, as shown in Chapter 6, Robert Artigiani has begun the potentially highly productive task of applying Prigogine's theory to a re-analysis of history, earlier the great laboratory for Weber and Pareto.[4]

The most widely known, large-scale body of what might be called pre-transformational theoretical work was carried out throughout the 1970s by various cross-disciplinary and systems-theory-oriented teams of investigators under the auspices of an unusual world industrial leadership initiative, the Club of Rome. The first of these studies was the well-known Limits to Growth study by Meadows and Meadows and associates, which used computer modeling to project the truly chaotic consequences if present trends within the prevailing global system continue into the next century.[43] Most significantly, in keeping with a stable versus instable state model for social analysis, this study also briefly examined the problem of social "chaos" against the background of the normative projection by philosophers such as John Stuart Mill of an ideally ordered no-growth, balanced or stationary state society.

Limits to Growth was followed by the Mankind at the Turning Point study by Mesarovic and Pestel, which in addition to refining a computer model of earth's functioning and projected future, again dealt with the idea that the global system is entering, or is well into, a form of chaos.[45] This study further advanced the normative thrust of the Club of Rome intervention by beginning to focus both on how to attain new social order out of this chaos and the ideal form for a new social order. Here the hard science oriented technologists of this study came up with a surprisingly "soft" answer. For they found that overriding all other

considerations—that is, more important than any specifics of economic
political, or technological change—was the need for a global systems-
wide change in *value* structure from "competition" to "cooperation,"
from "confrontation" to "partnership," changing our relation to nature
from "conquest" to "harmony," and basing all decisions not merely on
"short term considerations" but on "a sense of identification with future generations."[45,pp. 144-147,157]

A third Club of Rome study by systems philosopher Ervin Laszlo applied the systems theoretical perspective to dealing with the "goals for
mankind."[29] More recently, Laszlo has begun to put the Club of Rome
and "chaos" theoretical work within the larger perspective that provides
a home for both science and social action. This is the perspective of
general evolution theory, as articulated in Laszlo's concept of a Grand
Evolutionary Synthesis (GES).[33] The core dynamics of the transformational process in Laszlo's systems and cybernetical perspective condense
insights from many fields. They lie in the interaction of a "β-function
of auto- and cross-catalytic cycles governed by negative feedback
deviation-reducing mechanisms," in which systems governance is cen-
tralized, and a "δ-function of deviation-amplifying positive feedbacks,"
in which systems governance shifts to the periphery. Evolution is then
seen as the alternation of relatively long "epochs of stability" or "stable societal states, in which the β-function predominates" with a much
shorter "crucial epochs," or "revolutionary periods, characterized by
the δ-function."[30,pp. 141,145]

In keeping with the Club of Rome thrust, Laszlo calls for "interdisciplinary investigation" of societal transformation "by teams of
specialists sharing a common systemic and cybernetic orientation."[31]
He notes that the implementation of this research "constitutes one of
the most important challenges facing the contemporary scientific
community."[31,p.34] In keeping with the normative scientific heritage, he
projects that results "could include theories of practical guidance value
in the social sciences, reducing the uncertainty surrounding epochs of
rapid change by mapping out the possibilities and the dynamics of societal transformation and biasing the choices in favor of preferred
alternatives."[31,p.34]

This normative perception of both the fact of and the need for accelerating a fundamental shift in values lies at the core of physicist Fritjof
Capra's vast synthesis of "new paradigm" thought in *The Turning
Point*,[9] in the formation by Capra and others of the science-activist Elm-
wood Institute, and in the evolutionary approach to world problems of
biologist Jonas Salk.[57] Salk's concern, as Capra's, has been with the

fundamental problem for social disequilibrium theory of projecting possible kinds of order out of present disorder and how to "choose the most evolutionary advantageous path."[57,p.3] "Survival of the world as we know it is not possible," Salk concludes. "The world will have to be transformed and evolve for continued survival."[57,p.106] Similarly, Capra sees the necessity for "a transformation of unprecedented dimensions, a turning point for the planet as a whole."[69,p.16]

A critically important aspect of an adequate social equivalent to natural scientific chaos theory will be methodologies and data bases that make it possible to efficiently track change that may reveal transformational dynamics as well as potential intervention points and techniques. For well over a decade, psychologist Milton Rokeach and associates throughout many parts of the world have been carrying out experiments using Rokeach's Value Scale to explore value structural stable states. This work includes the induction, via cognitive dissonance, of the psychological equivalent of chaos states, and testing the effectiveness of interventions designed to produce fundamental and enduring values change.[56]

The most comprehensive and well-financed study of "chaotic" or transformational values change has been carried out over the past six years by California-based SRI International, one of the world's largest research institutions.[14] Basing its VALS (values and life styles) program on the work of psychologist Abraham Maslow, SRI social scientists regularly update a data bank on more than 100,000 Americans to predict changes in attitudes, values, and preferences that will affect the American and world economic, social and political systems.

In other SRI projects, a study by Duane Elgin of the problems of managing complex systems identified bifurcational options emerging from the state of systems breakdown—including further descent into chaos, authoritarian response, and transformational change.[20] A study by VALS research director Jay Ogilvy and noted political scientist James MacGregor Burns explores the relation of leadership styles to VALS findings[8] and Peter Schwartz and Ogilvy have explored the nature of the "emergent paradigm" in terms of "chaos" theoretical thought.[58] The present writer's work as a social psychologist has focused on the development of a psychologically adequate theory of directed social systems change and the nature of the human mind as an instrument for both predicting and shaping change. Eisler's work, as shown in Chapter 9, has addressed the need for a more complete data base for the study of cultural evolution and a new conceptual framework for both past and potential social systems change.

Working over a decade with participants in experimental groups, the author's research has defined and tracked the interactive dynamics of systems stability and change in terms of an exceptionally wide range of variables of demography, personality, ideology, and brain functioning. This work has empirically reduced change-stability-relevant variables to a primary handful, for which new measures and new theories have been developed to explain their interaction.[36,38] In particular, through the development of Ideological Matrix Prediction, this work has explored the dialectical impact of norm-changing and norm-maintaining as basic systems' orientations within individuals and groups. Both predicting and shaping the future have been found to be significantly linked to the relative thrust of these two polarities as further differentially expressed by the degree to which norm-changers or norm-maintainers are activists, extremists, tough or tender-minded, leaders or followers, and representing older or younger generations.[37]

The nature and function of the human mind within evolution has been further explored through the development of other tests and methods. These include new tests to define brain functioning in terms of left or right brain hemispheric dominance and interaction and a new non-sexually-biased test of moral sensitivity.[39,40] The centering perception for this work is of mind as an evolving *guidance* system for each person and for humanity as a whole. Research with people in a wide variety of settings indicates this higher guidance system includes the following set of frontal brain dependent "sensitivities": futures sensitivity (with an exploratory function), systems sensitivity (with an orienting function), moral sensitivity (with a directional function), social sensitivity (with both an orienting and directional function), and managerial sensitivity (with a decision-making function). Mind in this schema is further defined by viewing our time, the late 20th century, as a pivotal evolutionary dividing point between the dominance of the Truncated Mind and the crucial emergence and spread of Actualizing Mind, in which guidance sensitivities are more fully operative.[40]

Eisler's study of cultural evolution examines stability as well as oscillational and transformational change. Focusing on the fundamental organizational constraint of sex differences, it indicates that how the relationship between the two basic halves of humanity is structured has important implications for both the overall organization of social systems and human cultural evolution as a whole. Drawing upon a wide range of relatively neglected old as well as recent social scientific studies—in particular, recent and potentially revolutionary findings by archeologists in Anatolia, Crete, and Old Europe[44,51,22]—Eisler proposes two primary models of social organization characterized by widely differing social guidance or values systems. The first, designated the *partnership* or *gylanic* model (from the Greek roots, *gy* for woman, *an* for man, and *l* for their linking), is characterized by "soft" or stereotypically feminine values such as mutual accommodation, cooperation, and non-violence. The second model is the *dominator* or *androcratic* model (again from the Greek roots, *andros* or man and *cratos* for ruled), with a characterizing value and social guidance system idealizing "hard" or so-called masculine values such as conquest, mastery, and force.[17,18]

While recorded history mainly reflects variations of the dominator model, Eisler's study finds that both oscillational and transformational dynamics can be productively traced through history and prehistory in the enduring conflict between human agents responsive to one or the other model. Critical bifurcation points when major shifts in social structure and human consciousness have occurred, and may again occur, can also be recognized and predicted using this conceptual framework. From the perspective of Eisler's "cultural transformation" theory, we are now approaching a potentially decisive bifurcation point in our cultural evolution, with the centrally powerful androcratic or dominator paradigm driving us toward global tyranny and/or nuclear war—and possibly species extinction. By contrast, the gylanic or partnership paradigm, acting as the social equivalent of a "peripheral isolate" or "periodic attractor," is working to free the evolutionary potential of our species and transform the prevailing system before it self-destructs.

SUMMARY AND CONCLUSION

The magnitude of the problems humanity faces in these closing years of the 20th century, the investment over nearly 150 years of social research and thought moving in the direction of social theory adequate to handle disequilibrium as well as equilibrium processes, and the scope of contemporary work trying to move in this direction indicates the challenge and potential for a social equivalent to hard science "chaos" theory. Through informal networking many minds are already engaged in exploring this new territory for scientific and social advance. The hope we have tried to articulate, shared by many on both sides of the great watershed between natural and social science, is that by organizing a formal exploration of how "chaos" is to be understood and used for human advancement, we may avoid massive retrogression or annihilation.

The case for the application of "chaos" theory to the evolutionary challenge set forth here can be summarized as follows:

1. The founders of modern social science dreamed of a science that would at last provide humanity with a great tool to control its own destiny. Many noted analysts now conclude a central reason for the failure of social science to live up to this dream is because it is still mainly caught within the anti-normative paradigms of first and second stage scientific development. The social problems that confront us, however, are the third stage—the disequilibrium—reality. Hence, investment in developing third stage/normative social science is of the most critical, urgent and pressing importance.

2. The method of approach both explicit and implicit here is dialectical in its most ancient and enduring sense, transcending all modern ideologies. That is, projects to develop this theory would use social crisis as the generative force and real world laboratories for an examination of isomorphically relevant concepts in natural science, to in turn develop and advance relevant social theory, which in turn would be used to feed back *informed* humanistic guidance into social process.

3. Among the advantages of such an approach one is paramount: that it could provide a short cut to social payoff. Such a thought is alien to scientists and funding agencies accustomed to view science as, by necessity, a long, slow, careful process of the accumulation of knowledge over decades and, hopefully, centuries. But in the reality of this late 20th century world such an attitude seems a luxury that, in the particular area here in focus, humanity can no longer afford. Through sources ranging from the Club of Rome reports to daily newspaper and television accounts of starving in Africa, mounting missile emplacements, and a torrent of books documenting all the other crises of our time, the message is being drummed into us that time is running out. The overriding systems requirement for our times seems to be that we must abandon the philosophies of "play it safe," "go with the flow," or "business as usual," and find the will and the resources for social scientific interventions that are both swift and humanistic.

REFERENCES

1. Abraham, R., and Shaw, C. *Dynamics: The Geometry of Behavior, Parts I and II.* Santa Cruz, CA.: Aerial Press, 1984.

2. Abraham, R. *Brain/Mind Bulletin*, 9, 11/12, 1984, p. 2.

3. Allen, P. Self-Organization in Human Systems. In *Essays in Societal Systems Dynamics and Transportation Management 1981*. Washington, D.C.: U.S. Department of Transportation, 1981.

4. Artigiani, R. Revolution and Evolution: Testing Prigogine's Dissipative Structures Model. In Laszlo, E. (Ed) *The New Evolutionary Paradigm*. New York: Gordon & Breach, 1990.

5. Boulding, E. The Dynamics of Imaging Futures. *World Future Society Bulletin*, 7, 5, 1978, p. 2.

6. Boulding, K. *Ecodynamics: A New Theory of Societal Evolution*. Beverly Hills: Sage Publications, 1978.

7. *Brain/Mind Bulletin*, 9, 11/12, 1984.

8. Burns, J. M., and Ogilvy, J. A. Leadership and Values. Menlo Park, CA.: SRI International, 1984.

9. Capra, F. *The Turning Point*. New York: Simon & Schuster, 1982.

10. Csányi, V. General Theory of Evolution. *Acta. Biol. Acad. Sci. Hung.*, 31, 409–434, 1980.

11. Csányi, V., and Kampis, G. Autogenesis: Evolution of Replicative Systems. J. Theor. Biol., 114, 303–321.

12. Csányi, V. *Evolutionary Systems: A General Theory of Evolution*. Durham: Duke University Press, 1989.

13. Drucker, P. *The Age of Discontinuity: Guidelines to Our Changing Society*. New York: Harper & Row, 1969.

14. duBeth, D. Values and Life Styles. *Insight*, Bell Communications Research, 1, 15, September 19, 1984.

15. Durkheim, E. *The Rules of Sociological Method*. New York: Free Press, 1964.

16. Durkheim, E. *Suicide*. New York: Free Press, 1951.

17. Eisler, R. Gylany: The Balanced Future, *Futures*, 13, 6, December 1981, 499–507.

18. Eisler, R. *The Chalice and the Blade: Our History, Our Future*. San Francisco: Harper & Row, 1987.

19. Eldredge, N., and Gould, S. Punctuated Equilibria: An Alternative to Phyletic Gradualism. In *Models in Paleobiology*, Schopf, Ed. San Francisco, CA.: Freeman, Cooper, 1972.

20. Elgin, D. Limits to the Management of Large, Complex Systems. Menlo Park, CA.: SRI International, 1977.

21. Feigenbaum, M. Universal Behavior in Nonlinear Systems. Los Alamos Sci 1:4, 1980; see also Glieck below.

22. Gimbutas, M. *The Goddesses and Gods of Old Europe*. Berkeley, CA.: University of California Press, 1982.

23. Glieck, J. Solving the Mathematical Riddle of Chaos. *New York Times Magazine*, June 10, 1984.

24. Harman, W. *An Incomplete Guide to the Future*. San Francisco: San Francisco Book Company, 1976.

25. Hughes, H. *Consciousness and Society*. New York: Knopf, 1958.

26. *I Ching*. Princeton, N.J.: Princeton University Press, 1950.

27. Inkeles, A. *What is Sociology?* Englewood Cliffs, N.J.: Prentice-Hall, 1964.

28. Laszlo, E. *Introduction to Systems Philosophy: Toward a New Paradigm of Contemporary Thought*. New York: Gordon and Breach, 1972.

29. Laszlo, E. *Goals for Mankind: A Report to the Club of Rome*. New York: Dutton, 1977.

30. Laszlo, E. Cybernetics in an Evolving Social System. *Kybernetes,* 13, 1984, pp. 141–145.
31. Laszlo, E. Systems and Societies: The Basic Cybernetics of Evolution. Proceedings, 6th World Congress of Cybernetics and Systems, AFCET, Paris, 1984.
32. Laszlo, E. The Crucial Epoch. *Futures,* February 1985.
33. Laszlo, E. *Evolution: The Grand Synthesis.* Boston & London: New Science Library, Shambhala, 1987.
34. Lewin, K. *Field Theory in Social Science.* New York: Harper and Row, 1951, pp. 228, 229.
35. Lorenz, E. Irregularity: a Fundamental Property of the Atmosphere. *Tellus,* 36A, 98–110, 1984.
36. Loye, D. *The Leadership Passion: A Psychology of Ideology.* San Francisco: Jossey-Bass, 1977.
37. Loye, D. *The Knowable Future: A Psychology of Forecasting and Prophecy.* New York: Wiley-Interscience, 1978.
38. Loye, D. The Brain, The Mind, and the Future. *Technological Forecasting and Social Change,* 23, 267–280.
39. Loye, D. *The Psychology of Prediction.* Work in progress.
40. Loye, D. *Moral Sensitivity and the Guidance System of Higher Mind.* Work in progress.
41. Maslow, A. *The Psychology of Science.* Chicago: Regnery, 1966.
42. Maturana, H., and Varela, F. *Autopoiesis and Cognition: The Realization of the Living.* Boston: Reidel, 1980.
43. Meadows, D. H., Meadows, D. L., Randers, J., and Behrens, W. W. *The Limits to Growth.* New York: Universe, 1972.
44. Mellaart, J. *Catal Huyuk.* New York: McGraw-Hill, 1967.
45. Mesarovic, M., and Pestel, E. *Mankind at the Turning Point: The Second Report to the Club of Rome.* New York: Dutton, 1974.
46. Miller, G. *Living Systems.* New York: McGraw-Hill, 1978.
47. Mumford, L. *The Condition of Man.* New York: Harcourt Brace Jovanovich, 1944.
48. Myrdal, G. *Objectivity in Social Research.* New York: Pantheon, 1969.
49. Pareto, V. The Circulation of the Elites. In Parsons, T., Shils, E., Naegele, K., and Pitts, J., Eds. *Theories of Society.* New York: The Free Press, 1961.
50. Parsons, T. *The Structure of Social Action,* Vols. I, II. New York: The Free Press, 1968.
51. Platon, N. *Crete.* Geneva: Nagel, 1966.
52. Polak, F. *The Image of the Future.* San Francisco: Jossey-Bass/Elsevier, 1972.
53. Polanyi, K. *The Great Transformation.* Boston: Beacon Press, 1944.
54. Prigogine, I., and Stengers, I. *Order Out of Chaos.* New York: Bantam, 1984.
55. Riegel, K. *Foundations of Dialectical Psychology.* New York: Academic Press, 1979.
56. Rokeach, M. *The Nature of Human Values.* New York: Free Press, 1973.
57. Salk, J. *The Anatomy of Reality.* New York: Columbia University Press, 1983.
58. Schwartz, P., and Ogilvy, J. The Emergent Paradigm: Changing Patterns of Thought and Belief. Menlo Park, CA.: SRI International, 1979.
59. Sherif, M. *The Psychology of Social Norms.* New York: Harper & Row, 1966.
60. Sivard, R. *World Military and Social Expenditures* Washington, D.C.: World Priorities, 1983.
61. Skinner, Q. (Ed.) *The Return of Grand Theory in the Human Sciences.* New York: Cambridge University Press, 1985.
62. Sorokin, P. *Sociological Theories of Today.* New York: Harper & Row, 1966.

63. Sorokin, P. *The Crisis of Our Age*. New York: Dutton, 1941.
64. Spengler, O. *The Decline of the West*. New York: Knopf, 1932.
65. Thom, R. *Structural Stability and Morphogenesis*. Reading, Mass.: Benjamin, 1972.
66. Toffler, A. *Introduction to Order Out of Chaos*, by Prigogine, I., and Stengers, I. New York: Bantam, 1984.
67. Toynbee, A. *A Study of History*. New York: Oxford University Press, 1947.
68. von Bertalanffy, L. *General Systems Theory*, rev. edn. New York: George Braziller, 1968.
69. Weber, M. "The Social Psychology of the World's Religions." In Parsons, T., Shils, E., Naegele, K., and Pitts, J., Eds. *Theories of Society*. New York: The Free Press, 1961.
70. Weiner, N. *Cybernetics*. New York: Wiley, 1948.
71. Zeeman, C. *Catastrophe Theory*. Reading, Mass.: Addison-Wesley, 1977.

CHAPTER 3

Biological Evolution:
The Changing Image

GIANLUCA BOCCHI

Editor's Introduction: In the traditional and still dominant conception evolution is a biological concept, denoting the sequential, constant, and presumably gradual development of species from common primeval origins. This Darwinian idea is now seriously debated in regard to a number of particulars. The debates center on issues such as continuity vs. discontinuity, gradualism vs. saltatory dynamics, the role of chance and of necessity, and the importance of natural selection.

Bocchi reviews the main positions in these debates. The review throws light not only on the nature of biological evolution, the *prima facie* objective of the debates, but also on evolution as a general phenomenon. Especially noteworthy in the context of general theory are the emerging insights concerning the blending of indeterminacy and determinacy (or "chance and necessity") in the post-Darwinian—and also post-Monod conceptions, the discontinuities and leaps that intervene in the evolution of species, and the aleatory spread of mutants during processes of speciation.

As the various contributions to this volume testify, many of the new positions in the debates on biological evolution dovetail with conclusions that flow from independently researched fields. Bocchi's review of the state of the art in biology thus contributes to our understanding of the multiple strands of consistency that now emerge between fields as diverse as macroevolution theory, nonequilibrium thermodynamics, complex dynamical systems theory, learning theory, and the study of patterns of development in history.

CHANCE AND CONSTRAINT

Among the problems which have historically been of the greatest interest in the study of new evolutionary theories, the nature of novelty and creativity in biological evolution occupies an important place. This problem gave rise to sharp controversies among supporters of the Neo-Darwinian synthesis. In advancing the hypothesis that the already identified micro-evolutionary mechanisms and evolutionary mechanisms as a whole are identical, the Neo-Darwinian synthesis tended toward a mode of thought we can best characterize as "promissory." The identification of what was termed the central nucleus of the paradigm was held to suffice, during a longer or shorter interval, to thoroughly explain phenomena in all their multiplicity and variety. This created a substantial depreciation of the value of actual and future study, as epitomized in Jacques Monod's famous affirmation: "today we can affirm that the elementary mechanisms of evolution are not only comprehended in their general lines, but also identified with precision. . . . As far as the essential lines are concerned, the problem has been . . . resolved and evolution is no longer at the frontiers of knowledge" (Monod 1970, It. trans. 1970, p. 114).

Theodosius Dobzhansky, another of the fundamental masters of the Neo-Darwinian programme, offered a sharp criticism of this tenet. According to him Monod's point ignores the essential fact that evolution is a historical phenomenon, and that the laws which regulate it are not of a classical nature. "We do not yet have either a predictive or retrodictive theory of evolution. . . . Without palaeontological evidence, one cannot reconstruct the ancestry of the now living organisms. Brave attempts to do so were being made by zoologists and botanists, particularly of the late nineteenth and the early twentieth century, using inferences from comparative anatomy and embryology. The results were not satisfactory. . . . Given the path followed by the evolution of the human species, we may wonder if the *Australopithecus* was bound to evolve into *Homo erectus* and thus later into *Homo sapiens*. The question must, I believe, be answered in the negative. Given one stage, we could not predict what the next stage must be. There were at least two contemporary species of *Australopithecus*, but only one of them was our ancestor. The evolution of the other ended in extinction. Yet from our inability to make such prediction it does not follow that the human species arose by a lucky throw of some evolutionary or celestial dice. We neither arose by accident nor were predestined to arise. In evolution, 'chance and destiny' are not alternatives" (Dobzhansky 1974, pp. 328–29).

Here is delineated a mode of thought which, though based on the acknowledgement of the *same* "facts" and the *same* evolutionary factors, is considerably different from many interpretations current to Neo-Darwinism. A crucial difference arises between one who considers evolution as a product of a collision between chance and necessity, of relatively dissociable and analyzable factors, and one who emphasizes their interactions—along with the constructive character of this interaction—within the sphere of complex systems of interactions and regulations. Jacques Monod's study was based precisely on a hypothesis of the separability of the function of chance and of necessity. Generally speaking, an innovative function in evolution was ascribed to chance which presided over mutations, whereas necessity, which belonged to the teleonomic reproduction apparatus, constituted a strongly conservative factor. This form of explanation proved to be closely integrated with the epistemological assumptions characteristic of classical science, according to which a thoroughly scientific explanation is given only of the invariant and necessary aspects of phenomena. On the contrary, chance and contingent events are generally considered irrational and are the object of strategies which tend to belittle their importance for scientific explanation. It appears that the central aspect of a theory of biological evolution is furnished by its synchronous and invariant aspects, belonging to the teleonomic reproduction apparatus which is explainable in chemical and molecular terms. On the other hand, its truly historical aspects, considered as ''specific differences'' within a framework of substantially given invariants, are left in shadow. It is therefore comprehensible why we insist on the complete and thorough knowledge of evolutionary mechanisms, emphasizing more what has been explained than what must still be explained.

This is not the only attitude possible, though. We can assume, on the contrary, the *fact* of evolution's creative character as the point of departure, as it is manifested in the production of new forms (though on the basis of the combination of preexisting elements), and we can seek an explanation for it by attributing to necessity, too, a formative, and not simply a conservative character. This calls for surpassing the classical concepts of scientific laws and causal determination of a form that is analogous to, though it is vaster than that which has existed in physics for many decades. The problem does not lie in the relationship between an innate necessity in laws which indicate forever constant and equally determined processes and pure chance to which laws cannot be applied, but rather in the relationship between the *constraints* which every ''historical'' situation imposes on evolutionary processes (constraints which, in this particular instance, derive from biochemical mechanisms or, more specifically, from the characteristics of the level of organization of a given species) and the *possibilities* compatible with, and therefore opened up by, these constraints. If pure chance *can* preside over the *realization* of a certain possibility in place of another, we cannot say that its *existence* is completely indeterminate.[1] But this determination is plurivocal and not univocal and can be known only *a posteriori* when the possibility has become actualized. In this perspective, we can envisage the possibility of an evolution of the laws and of the evolutionary constraints themselves. Some constraints can rise, decisively channelling the course of evolution: this is the case, for example, of the genetic code whose universality and necessity for all organisms today living was most probably preceded by a long period of trial and error during which embryos with alternative codes were probably produced. In the same way, other constraints may become attenuated or change in significance: the quantity of information which can be stored in an organism is always limited, but it is one thing to utilize only the genome for storage, and another to utilize the greater capacity and flexibility of the nervous system. The constraints do not disappear, but rather change in time and place.

The explanatory schemes derived from classical physics and from its atemporal characteristics are inadequate for the theory of evolution. Today, evolutionary theory takes the form of a truly historical science with a specific rationality and explanatory system. If Dobzhansky emphasizes the fact that a predictive theory of evolution could not exist yet, the accent has progressively moved, in the course of the seventies and eighties, to focus on the fact that such a theory, founded on the "classical" concept of prediction, simply cannot exist. Its impossibility is intended not in relation to our ignorance, but in relation to the profound mechanisms of interaction between constraint and contingent factors which characterize evolution, inasmuch as it is a historical phenomenon. We are faced with a change of epistemological model within which evolutionary theories can be constructed; in more general terms, we face a change in the notion of prediction. Steven Stanley, one of the greatest contemporary palaeontologists, writes: "If a being gifted with perception had observed, without intervening, human evolution in the three or four million years preceding the appearance of *Homo sapiens* and his immediate ancestors, he would not have been able to foresee our origin. Our anatomy and the very different one of our thickset Neanderthal cousins are, in many ways, unforeseeably departing from everything which preceded them. Our body functions well enough, but no better than other combinations of flesh and bones, which could have evolved in our place from *Homo erectus,* our father and grandfather from the evolutionary viewpoint. The sharp and jutting chin, high forehead and rather rough superciliary ridge are not the result of evolutionary tendencies which operate on a long term; they are, instead, unique characteristics of our species. They represent, in evolution, true surprises even though they may have been favoured by natural selection in a particular moment and in a particular place" (Stanley 1981, It. trans. 1982, pp. 13–14).

CONTINUITY AND DISCONTINUITY

The debate as to the constructive character of evolution appears closely connected with the continuist or discontinuist debates that occurred from time to time. As Richard C. Lewontin and Richard Levins have pointed out, one viewpoint, in particular, which tends to devaluate the constructive character of evolution and to overestimate its recombining character seems closely connected with the origins of the Darwinian programme itself. It arose with the intent of giving a continuist explanation to a macroscopic reality which appeared (as Cuvier had pointed out) to be decisively discontinuous and which could not be treated with the conceptual heritage of the times. "Discontinuous or qualitative change remained . . . mysterious. Convinced that 'nature does not advance by leaps,' Darwin appealed to the existence of intermediate forms to corroborate his theory, admitting the gravity of the absence of intermediates or of the incompleteness of fossil finds. Successively, biologists, forced to accept the evidence of evolution by transmission of modifications, sought the way to attribute a mere epiphenomenical value to evolutionary change. Hence was born August Weissmann's idea for which the rich diversity of animal life was *only* the recombination of a hypothetical invariable 'ideoplasm.' On the other hand, many modern efforts to define evolution as a variation of the gene frequencies in populations reflect the prejudice that qualitative constancy is more fundamental than change, that qualitative differences are, in a certain sense, illusory. However, science has reluctantly accepted the reality of qualitative change and the importance of discontinuity. Familiar examples are the transitions in phase between the solid, liquid and gaseous states" (Lewontin/Levins 1978, pp. 1027–28).

The problem of evolutionary innovations can be adequately tackled by taking note of the non-simple and non-univocal nature of the interactions which determine evolutionary processes. However isolable it may prove in the abstract, any element subjected to an evolutionary process (be it a chromosome, an organ or a morphological or behavioural "trait") will always evolve following the pushes and constraints provoked by the elements with which it interacts, as well as those provoked by the broadest environment of which it is part (be they of a cellular or ecological order). The result is a kind of course which is not at all linear, but rather full of oscillations and apparent reversions. This is what is commonly referred to as the *opportunism* of evolution, expressed in François Jacob's well-known metaphor according to which evolution does not work like an architect or an engineer who follows a pre-arranged plan, but rather like a *bricoleur* who must assemble objects which come from many different places, whose original functions often have nothing to do with the purposes he assigns to them.[2] In analyzing these processes as they take place in the evolution of organisms, which emerges is the extreme variety of the ways in which a single function can come about according to evolutionary contingencies and the available material. Once again, this induces us not to trust in predictive (or retrodictive) power (in the strict sense of the word) of the theory of evolution, and instead to conceive of processes characterized by numerous possible bifurcations of which only a few are realized. "All living beings have basically the same requirements—food, place to live, reproduction, protection from environmental insults. Yet there are millions of species that solve these requirements in million of ways. To give more specific examples—a warm-blooded animal can survive cold winters by having a warm coat, by winter dormancy, by migration to warmer lands or by using a fire to heat properly constructed dwellings. A desert plant may protect itself from desiccation—by having leaves turned into spines, by leaves covered with oily or resinous varnish, leaves that develop during wet seasons and are shed during dry ones, or by compression of the life cycle to fit the wet season . . . " (Dobzhansky 1974, p. 331).

THE EVOLUTION OF EVOLUTION

The scheme of constraint and possibility also brings to the forefront the problem of the mutation of the constraints and the laws which preside over the channeling of evolutionary phenomena; the problem of the evolution of the evolutionary possibilities themselves. This aggravates further the inadequacy of any inventory of evolutionary possibilities claimed to be definitive and thorough, and to determine a rigidly set space within which all possible actualizations of terrestrial organisms, or of a given type of organization, can be set. The boundaries of such spaces seem, instead, to be entangled and, at least in some cases, subjected to sudden and unforeseeable redefinitions. Lewontin and Levins are very clear in affirming this: " . . . every trait presents additional properties which join those initially selected. These properties (the non-selected consequences of selection) create possibilities but also new vulnerabilities and, with a change in the circumstances, can become the principal store of selection. . . . "

"In the extreme case, the impossible first becomes possible and then necessary. The most probatory example was the revolution of oxygen. Oxygen is a very toxic substance for organic matter and, a long time ago, the capacity of protecting oneself from oxygen or of avoiding it must have had an enormous selective value. Still today, there are anaerobic organisms which survive by installing themselves in places where oxygen cannot penetrate. Certain organisms, though, confronted oxygen in a different manner: they deprived it of its toxicity by allowing (in practice, by favouring) the interaction with cellular organic substances. In this manner, not only did the oxygen molecules lose their poisonous characteristics, but they could be exploited to release accumulated chemical energy. Metabolic efficiency thus drastically increased and permitted aerobic organisms to acquire a position of absolute pre-eminence among living beings. Finally, the dependence upon oxygen gave rise to new vulnerabilities. Indeed, the lack of oxygen constitutes a more immediate threat for life than the absence of food; the majority of organisms are excluded from habitats with a poor supply of oxygen; various internal organs developed to distribute oxygen efficiently, and the conditions which prevent this distribution (circulatory problems, anemia, carbon monoxide poisoning) constitute new threats to survival'' (Lewontin/Levins, 1978, pp. 1029–30).

The production of new evolutionary forms and possibilities depends essentially on the network of interactions which exists among the constituent elements of a given organism, between one organism and others, between organisms and the environments of which they are part, and between the environments themselves. These interactions are not logical operations which disassemble and recompose a thoroughly defined dominion outside of specific spatial-temporal conditions. Rather, they are a series of circular and retroactive causal chains endowed with properties of self-equilibration and self-amplification, and characterized by the attributes of historicity and singularity. Faced, for example, with the Earth's ecological condition two billion years ago, with an atmosphere extremely different from that of today, an external observer could easily hypothesize the abstract possibility of aerobic organisms for which oxygen no longer constitutes a poison. Even for such an observer, however, it would be impossible to foresee and understand *why* these organisms, which were originally extremely deviant, were successful in spreading to the point of almost complete dominance over the planet.

The epistemologies of the classical school have considered the incapability of producing linear predictive chains as a function of ignorance, maintaining the possibility of a thorough combinatorial and perfectly foreseeable regulatory ideal, a kind of "demon of Laplace." In this sense, the epigenetic and creative aspect of evolution would be epiphenomenal, founded on more basic performist mechanisms. The introduction of the concepts of interaction and retroaction transforms the concept of prediction in scientific explanation, producing an image of reality—of evolutionary reality in particular—which is coherently epigenetic, creative and self-transcendent.[3] Post-Darwinian modes of thought are in the heart of this transformation.

GRADUALIST AND SALTATORY EVOLUTION

The contemporary transformation of evolutionist thought shares the here outlined fundamental epistemological framework. It is on this basis that the classical Darwinian assumptions are investigated. In this context, the still poignant controversy over the evolutionary rhythms of biological organisms has a special importance. In general outlines, two positions confront one another: the gradualist and the saltationist. The former, with its strict Darwinian orthodoxy, conceives of evolution as the product of a series of slow and imperceptible mutations accumulated almost exclusively through the effect of natural selection. The latter, the saltationist, produces an image of natural history as characterized by long periods of stability (in which natural selection acts as a preserving force) and by periods of crises, fluctuations and bifurcations which, though brief on a geological scale, are decisive for the production of innovations in biological evolution (at least of the most important ones such as new species or new genera). The saltationist hypothesis was advanced in the seventies by several biologists such as Stephen Gould, Niles Eldredge, Steven Stanley and Richard Lewontin. They emphasized how the greatest evolutionary innovations on the phenotypical level are produced in the course of the speciation processes, how these processes depend on a multiplicity of factors in which the role of natural selection is sometimes rather tenuous if not null, and how the process of speciation is almost instantaneous if compared with the duration of a species.

The saltationist outlook expresses a hypothesis concerning the nature of evolutionary change, to be corroborated or falsified by means of adequate empirical checks (be they produced by data from palaeontology, genetics, or from elsewhere). The studies in the field of biological evolution in recent years have discussed the degree of corroboration or falsification of saltationist hypotheses. The questions posed are of the type: Is it possible to elaborate a coherent image of natural history of the saltationist type? Are hypotheses of the saltationist type perhaps not necessary to explain the great moments of natural history (such as the explosion of the Cambrian period and the origin of man)? And can these events have as plausible an explanation in saltationist terms? In any case, what is the relative frequency (quantitative and qualitative) of the intervention of mechanisms of a saltationist order in the course of natural history?

The two positions at stake are divided not only by the acceptance or rejection of certain scientific hypotheses subject to an empirical check. What is in question is also and above all the epistemological framework of the classical assumptions of Darwinism and Neo-Darwinism. The debate on gradualism is only an element in a more general effort of critical analysis of the concepts and ways of thinking inherent to the Darwinian tradition, aimed at their relativization and more adequate utilization in real evolutionary processes. This is evident in several writings in which Gould presents the most significant features of the turning points of contemporary evolutionism. In more general terms, "two sorts of proposals" are advanced: "(1) a widened role for non-adaptation and for chance as a source of evolutionary change, including claims that non-selected features act as important channels for pathways of change (though selection may propel organisms down the permitted paths), and as facilitators in forming pools of co-optable features for future change; (2) attempts to construct a hierarchical theory based on the interaction of selective (and other) forces at numerous levels (from genes to clades) rather than almost exclusively upon selection among organisms" (Gould, 1983, p. 352).

Through these formulations, the saltationist theories of contemporary evolution—or theories of punctuated equilibria, as they have been defined—lead to evolutionary pluralism. This constitutes an important attempt to elaborate a science of change outside the epistemological framework regulated by the concept of a fundamental point of explanation. The idea that there exists a privileged and unique rhythm of change fails, and with it the idea of homogeneous change which could justify the explanation of the temporal by the atemporal. In a certain sense, two distinct notions of change appear, referring to different rhythms and to different logics of change. Their interaction is complex and truly constructive of the multiform and variegated image of natural history. Within this epistemological model the idea of a new unitary theory capable of rendering the rhythms and logics of change homogeneous is not plausible. The rhythms and logics of *intraspecific* evolution are essentially stabilizing and conservative, and the fluctuations which operate in this framework tend to be attenuated in the vicinity of the particular equilibrium points which a given species sets itself from time to time. The classical Darwinian concept of natural selection and the slow accumulating of small mutations can often prove adequate for this concept of change, relative to long or extremely long periods in the history of a species; and the constraints imposed by the stability of the various defining equilibria of the species can be such as to give the concept an almost determinist nature in the evolutionary process (chance does not have a truly constructive role, given that the fluctuations are attenuated). The logic of *interspecific* evolution, however, is completely different. It is characterized precisely by the amplification of the fluctuations, by the possible constructive and decisive influence of singular and contingent events, by the slackening and transformation of certain constraints and the constitution of new points of equilibrium. The concepts developed within the Darwinian tradition, up to the Neo-Darwinist synthesis, were subjected to an epistemological model which believed that the various rhythms of evolution can be seen as homogeneous, and its logic as unitary. The debate presently underway has shed light on the inadequacy of the traditional conceptual store and on the necessity for its reinterpretation within a new epistemological model. Old questions shift, change meaning, others simply lose sense. As an example of these epistemological changes, we shall analyze in greater detail the concepts of "natural selection" and "adaptation," and the debate on the privileged nature of their relationship (which in the past had seemed natural).

THE ROLE OF NATURAL SELECTION

Besides the *dynamic* description of evolutionary factors, which may be summarized in the three principles of variation, heredity and natural selection, Darwin had also proposed their *kinematics* closely related, as is well-known, to the ideas of Malthus. Variations among individuals proved significant because individuals were considered to be immersed in an ecology provided with an unsatisfactory quantity of resources in comparison to the proliferation of living species. The individuals which survived and spread were those which possessed traits more suited to certain environmental conditions, and this process of progressive adjustment to an everchanging environment was defined as *adaptation*. Darwin presented this framework as fundamental but not as thorough. Significant in this regard is a passage which he added to the last edition of *The Origin of the Species* in 1872: "But, as my conclusion have lately been much misrepresented, and it has been stated that I attribute the modification of the species exclusively to natural selection, I may be permitted to remark that in the first edition of this work, and subsequently I placed in a most conspicuous position—namely at the close of the Introduction—the following words: "I am convinced that natural selection has been the main, but not the exclusive means of modification." This has been of no avail. Great is the power of steady misrepresentation." (Darwin 1872, ed 1952, p. 239). Darwinian tradition, however, has often developed precisely by these misunderstandings. By the end of the 19th century the concept of natural selection became an omnipotent explanation of all evolutionary phenomena, and Neo-Darwinism remained substantially within this framework. Contemporary evolutionism now seeks a revision of this assumption.

A series of indications leads to the belief that the phenomena of speciation, the importance of which we have already emphasized for macroevolution, involve a slackening rather than a strengthening of selection. "If one were to accept that competition in some way 'caused' the genetic divergence of some forms, and that sufficient genetic divergence would bring about speciation; then one might expect the most abundant speciation to occur within the same population in the same geographic area (sympatric speciation). This was in fact what Darwin (1872) assumed, though evidence was accumulated even then that geographical isolation almost always preceded speciation. . . . The modern view, due chiefly to Mayr (1954), is that isolation in one form or another is a prerequisite to speciation. . . . Regardless of whether one finds Mayr's genetic explanation acceptable, *isolation* certainly figures prominently in most evolutionary changes. This in itself considerably reduces the role of natural selection. In fact, one cannot but notice that the really conspicuous factor in all cases of rapid evolutionary change is the relaxation of natural selection. Could it be that the rapidity of changes observed is due as much to the *lack* of competition as to 'genetic revolution'?" (Ho/Saunders, 1979, pp. 577–78).

The omnipresence of natural selection as a factor of evolutionary mutation is also an issue in the so-called *neutralist* conception which, in the course of the seventies, was advanced especially by Motoo Kimura. No morphological or physiological trait possesses of itself a positive or negative evolutionary value. This depends instead on the combinations of the whole (of the genome or the organism) of which it is part, as well as on the environmental circumstances. Nonetheless, the classical Darwinist and Neo-Darwinist programmes maintain that the reasons for the diffusion (or for the failure) of a given trait always depend, in some way, on the value it assumes with respect to the pressures of natural selection at a specific time and in a specific environment. New developments in genetics have made us familiar with the notion of the polymorphism of a certain trait, that is to say the presence, in a given population, of different varieties of a trait. From the viewpoint which tends to accentuate the role of natural selection, the polymorphism of a given trait is taken as a rare or, in any event, as a transitory situation, a passing phase which accompanies the diffusion and the evolutionary success of a new variety. Recent studies on proteid evolution, though, show that the contrary may be true. Kimura maintains that many mutants are spread simply because they are neutral: "the majority of mutants which spread in the species are neutral, even though the neutral mutants constitute a small fraction of the mutants which are produced on the whole in a given moment. Those mutants which are destined to spread throughout the species require a long period before becoming established, and along the way they assume the form of a 'proteid polymorphism' " (Kimura/Ohta, 1973, p. 469). In this case, polymorphism appears as the norm, even though it may be supposed that it is attenuated in periods of accentuated selective pressure. A problem arises then which proves to be perfectly parallel to the one posed by the modalities of speciation: it may be that an excessive selective pressure is an obstacle precisely because it sets limits to a polymorphism of traits within which the production of evolutionary innovations would be favoured.

Even if the possible epistemological interpretations of neutralism vary, in any case they have furnished an important contribution to the development of the pluralist thinking underlying contemporary theories of evolution. By underlining the importance of factors extraneous to natural selection (in this particular instance, chance and drift) in evolutionary processes, they debate a true dogma in the Darwinian programme: *that all traits are in some way adaptational,* a punctual expression of the organism's answers to specific environmental necessities.[4] This assertion is not only disputable on the empirical level; it also involves great methodological problems. Moreover, it may be used in a completely arbitrary manner if generalized to behavioural or even psychological fields.[5] "In adaptational analysis are concealed numerous assumptions which go back to the theistic conceptions of nature and to a naive Cartesianism. In the first place, it is necessary to depart from the assumptions that the division of organisms into traits and that of the environment into problems have a real basis and not one which is simply the materialization of intuitive human categories. In what sense in nature is a fin, a leg or a wing an individual trait whose evolution is comprehensible in terms of the solution of a particular problem? If the leg is a trait, are each of its parts as well? At what level of subdivision do the confines no longer correspond to 'natural' divisions? Perhaps the entire topology is erroneous. For example, the normal physical divisions of the brain correspond quite approximately to the localization of certain central nervous functions, but it seems that the memory of events is accumulated on a widespread basis, therefore a certain microscopic region is not the seat of particular memories. If we move from anatomical characteristics to descriptions of behaviour, the danger of materializing human categories increases. Animal behaviour is described by means of categories such as aggression, altruism, parental cares, war, slavery, cooperation and for each of these 'organs of behaviour,' an adaptational explanation is furnished by identifying the problem for which it represents a solution" (Lewontin, 1977, p. 204). The weakest point of an exclusively adaptationist explanation lies in the fact that it risks unduly multiplying the recourse to *ad hoc* hypotheses. Faced with an evolutionary problem, a particular trait is hypothesized and isolated more or less arbitrarily, and its adaptational development is made to explain it satisfactorily. If things do not function, another trait and another explanatory chain are sought, and so on according to a series which progressively diminishes the explanation's degree of controllability and increases its complexity.[6] Those who declare themselves explicitly Anti-Darwinist are in a good position to declare a continuist explanation as completely on the wrong track in regard to small variations, and to search for the heart of evolutionary mutation in "internal factors," unknown to date.

However, the mutation to be explained may be of a different type. That is to say, it is not a question of moving from one univocal and centered explanation to another explanation as univocal and centered as the first, but rather to put explicitly on the table the question of proliferation in evolutionary factors and mechanisms. Though for many phenomena the traditional Darwinian explanation conserves its validity, this does not mean that extrapolation from this to the totality of phenomena will always furnish good results. In place of a single mechanism which repeats itself identically in all evolutionary processes, it may be that evolution creates a complex typology. There may exist a variety of ways in which selection and adaptation can combine in evolutionary mechanisms, besides the classical Darwinian one. Gould and Lewontin speak[7] of:

1) An absence of selection and adaptation. This is the case of casual drifts which may be connected with the question of neutralism;

2) An absence of selection and adaptation as far as the trait directly in question is concerned; this trait may, instead, owe its form and evolution to an effect resulting from direct selection processes operating elsewhere (on other traits);

3) A rupture of the privileged relationship between selection and adaptation both in the sense of selection without adaptation and in the sense of adaptation without selection;

4) Adaptation and selection without, however, the latter furnishing criteria for deciding which is the best among different types of adaptation;

5) Adaptation and selection according to a particularly opportunist adaptational strategy which utilizes only in a secondary manner the materials originally present for general structural reasons or due to prior evolutionary history.[8]

Within the development of the Darwinian tradition, a controversy occurs between a pluralist and a monist explanation the importance of which goes beyond purely evolutionary phenomena and concerns all of biological science: "We feel that the potential rewards of abandoning exclusive focus on the adaptationist programme are great indeed . . . . non-adaptative does not mean non-intelligible. We welcome the richness that a pluralistic approach, so akin to Darwin's spirit, can provide. Under the adaptationist programme, the great historical themes of developmental morphology and *Bauplan* were largely abandoned; for if selection can break any correlation and optimize parts separately, then an organism's integration counts for little. Too often, the adaptationist programme gave us an evolutionary biology of parts and of genes, but not of organisms. It assumed that all transitions could occur step by step and underrated the importance of integrated developmental blocks and pervasive constraints of history and architecture. A pluralistic view could put organisms, with all their recalcitrant, yet intelligible complexity, back into evolutionary theory." (Gould/Lewontin, 1979, p. 597).

THE NEW EPISTEMOLOGY OF EVOLUTIONARY THEORIES

Contemporary biology (and evolutionism in particular) helps to rediscover a line of thought based on the concepts of form, organization and organism which can be traced back to Goethe and which, in the twentieth century, has found adherents in Paul Weiss, Ludwig von Bertalanffy and Rupert Riedl. This is not necessarily contrary to Darwinism, but can constitute its completion and the source of new creative ideas. Particularly important is the way in which the new approach deals with the question of constraints. It maintains that the variability of the part subjected to evolutionary pressure is not infinite, because it is dependent on the total organization of the morphological, physiological and behavioural systems in which they are inserted, as well as on the environmental demands and requisites manifested under the form of selection. In this way a great circle is created between organism and environment which may have different manifestations and mechanisms. In it the organism can even perform an active role and transform its ecology.

A reference to D'Arcy Thompson is indicated here. In the twentieth century he stood, in a certain sense, as the great heir of Cuvier, or rather as the one who, faced with Darwinian continuism, insisted the most on the discontinuity, around a few fundamental forms, of the *scala naturae*. He attempted to show how the various forms of living organisms are transformations of these general forms (which may be described, at least in part, in mathematical terms) and how biological variability encounters major constraints in purely physical requisites (for example, the maximum dimensions of a type of organization within which it can remain coherent). The small attraction of this approach for evolutionists is understandable, especially if they are convinced of the omnipotence of selective pressure and the continuity between the microscopic and the macroscopic aspects of evolution. In any case, D'Arcy Thompson's approach to the origin of the various forms of organization was considered vague. He attributed their origins to macromutations about which, at his time, not much could be said. Today, however, the questions he left open can be envisaged in a more considered and fertile way. Discontinuities in many aspects of evolutionary phenomena (the "missing links" which are missing not because they cannot be found, but because they have never existed) need to be explained even by Neo-Darwinists. At the same time we begin to understand how mutation itself is a stratified phenomenon which cannot be reduced to simple changes in amino-acids within a proteid structure.[9] Molecular biology has shown how the development of the organism is a complex network of relationships in which an essential role is played by some regulatory genes which do not codify any particular trait, but rather the sequence of the activation or deactivation of the genes themselves. Evolutionary leaps seem to be connected with the organism's superior level of organization: " . . . major anatomical changes usually result from mutations affecting the expression of genes. According to this hypothesis, small differences in the time of activation or in the level of activity of a single gene could in principle influence considerably the systems controlling embryonic development. The organic differences between chimpanzees and man would then result chiefly from genetic changes in a few regulatory systems, while amino-acids substitution in general would rarely be a key factor in major adaptive shifts" (King/Wilson, p. 114).

The origin of the giant panda *(Ailuropoda melanoleuca)* constitutes a good indication of the importance of the regulatory gene for macroevolutionary mutations, and it is almost a symbol of the change in viewpoint provoked by the theories of evolution of the last twenty years. According to traditional taxonomic criteria, there was doubt as to whether the giant panda should be connected to the bear family or to the lesser panda and, thus, to the family of raccoons. D. D. Davis in 1964 showed that the panda belongs to the family of the bear, though in a manner so aberrant as to constitute a subfamily with a single genus (or even a family related to the *Ursidae* for genetic reasons). The aberrant aspect lies in the vegetarian nature of the panda which, in comparison to the bear, involves a series of very evident morphological changes connected with its type of nutrition (teeth, masticatory musculature, mandibulary articulation). And yet, subjected to immunological analysis, the panda revealed a biochemical structure quite similar to that of the bear, with a variation analogous to that which exists between other two closely related species, such as the dog and the fox. The hypothesis is that the panda and the bear differ because of a very limited series of biochemical mutations (perhaps only in four or five *loci*), articulated, however, in such a manner as to influence decisions situated towards the top of the organism's hierarchical scale of development and therefore influencing a multiplicity of morphological traits by the "cascade effect." Steven Stanley rightly considers the case of the panda as one of the best examples of quantum speciation. The traits are so closely related as to render the application of a gradual scheme of reciprocal coevolution simply impossible.[10]

It is the origin of man, however, which shows the full import of the evolutionary role of these mechanisms. In the already cited article, King and Wilson present the results of studies on the evolutionary distance between man and chimpanzee. From the biochemical viewpoint the two species prove to be surprisingly similar: more similar than those zoologists call "sister species" (having almost identical morphological traits), and also more similar than the average two species belonging to the same genus. This, evidently does not cancel the extreme distance (not only behavioural and intellectual, but morphological as well) which exists between man and chimpanzee. It does mean, however, a new hypothesis of the origin of the difference. Instead of deriving it from numerous evolutionary variations, it may be traced back to a very few quantic events and their consequences. Stephen Gould suggested that one of the decisive events, if not *the* decisive one, is a foetalization process of man with respect to his forebearers.[11] It has been known for some time that the newborn human in the first phases of life is a sort of "extra-uterine embryo": his development has not yet reached its conclusion. A comparative analysis of the human infant with the newborn of other primates shows him in a very unfinished state. This is a case of *neoteny:* the maintenance by adult organisms of what were once only youthful traits following a delay in somatic development.

In the development of the human species neoteny may have had a primary adaptational value. Childbirth among the primates is a difficult operation: giving birth to a newborn infant not yet fully developed would have the advantage of allowing the growth of his brain beyond an otherwise unreachable point. Besides having direct consequences on a series of morphological traits, the neotenic nature of the human species may have a direct link with his more evident specificity: the enormous increase of cerebral capacity, with all its consequences for learning and intellective capacity. The neotenic hypothesis of the origin of the human species promises to unify in a coherent framework a series of otherwise dissociated elements. It is accompanied by a discontinuist interpretation of the origin of neotenic tendencies. Neoteny is, in fact, part of the extremely vast sphere of heterochrony phenomena, which indicate drifts and changes in the course of the ontogenesis of an organism. It is plausible that all heterochronous phenomena depend in various ways on mutations within the regulatory apparatus which, for example, would modify the temporal order of activation of certain genes, invert some priorities, slow down or accelerate the rhythms of epigenesis, etc. Seen from this viewpoint, the origin of the human species would become the most important case of quantum speciation, destined to a great future for his adaptational value as well as for the capacity to catalyze further transformation processes which have made *Homo sapiens* diverge more and more from his close relatives. This hypothesis is indirectly confirmed by a series of studies on the various types of heterochrony phenomena which tends to see it as an important source of evolutionary mutation.

The approach of punctuated equilibria does not constitute a depreciation of biochemical differences in favour of morphological differences, but rather an effort to relate them. The search of the causes of macroevolution leads to a close examination of the organism's genetic, biochemical and developmental mechanisms. The principal ally of the theory of evolution today is embryology. A systemic and stratified conception of the organism surfaces as the fundamental framework for such a connection: "A biological system is composed of many hierarchical levels or subsystems, and indeed interlocking hierarchies. In bacteria, among the simplest of all biological systems, an operon is commonly controlled not only by its own regulatory gene but also by 'higher' level systems which determine the states of many otherwise unrelated operons simultaneously. . . . At the global level, therefore, the system may be described by a *configuration* of states in its component subsystems. It is quite possible that alternative *stable* configurations exist and that transitions between configurations may be governed by the same sort of thresholds that determine state transitions in each subsystem. A change in the order parameter at some sufficiently 'high' level would automatically initiate a cascade of changes at the lower levels and eventually lead to the stabilization of the system at an alternative configuration. This rich dynamical structure thus embodies both stability and organized change." (Ho/Saunders, 1979, p. 584).

This model draws its origins from the studies of Conrad Waddington on canalization, chreod and homeorhesis. Waddington was among the first to emphasize that the real objects of natural selection are not the abstract genotypes, but the phenotypes produced by the organism's epigenetic process of development.[12] This process exhibits traits of singularity and historicity, in the sense that it is the product of an actual interaction between genetic messages and disturbances and environmental circumstances; it is not the result of immutatable rules. It also exhibits strong stability traits with respect to the variety of possible disturbances. Epigenetic development is homeorhetic, that is to say, it tends towards producing results that are quite similar even in different environmental conditions. Embryological research conducted by Waddington has shown that the stability trait produces discontinuous change: the passage from one epigenetic course *(chreod)* to another does not take place gradually, but by "catastrophes." Precisely in considering how stability is an essential property of all evolutionary systems, we understand in what way *heterorhesis*—the passage from one chreod to another, from one form of stability to another—can constitute one of the fundamental factors of macroevolutionary change: " . . . internal thresholds are capable of stabilizing a course of development. This stabilization takes place within a certain sphere of environmental conditions which we can define as the *subthreshold* domain. This implicates that the variations outside of this *(above the threshold)* have, as result, the cessation of homeorhesis. When this happens, it does not necessarily mean cessation of development or death; on the contrary, the system in course of development in the end may develop into an alternative global configuration by following some potential ramification of the course of development. It could, that is, furnish a *heterorhetic* response which considerably deviates from the normal phenotype, be it that it prove directly adaptational with respect to environmental stimulus, as seen in the case of various stabilized phenotypical polymorphisms . . . , be it that it otherwise prove all the same compatible in the first instance with life." (Ho/Saunders, 1979, pp. 585–86).

Francois Jacob introduced the notion of *molecular bricolage* to explain this new evolutionary dimension. Molecular bricolage is reducible neither to great mutations in the biochemistry of organisms (which by now, in all probability, are excluded from natural history), nor even to the punctiform mutations of gradualism and the synthesis of the past decades. Molecular bricolage constitutes the evolution of organization and of regulation: "What distinguishes a butterfly from a lion, a hen from a fly, or a worm from a whale is much less difference in chemical constituents than in the organization and distribution of these constituents. The few really big steps in evolution clearly required the acquisition of new information. But specialization and diversification took place by utilizing differently the same structural information. In the vertebrates, chemistry is the same. As has been often emphasized, differences between vertebrates are a matter of regulation rather than of structure. Minor changes in the regulatory circuits during the development of the embryo can deeply affect the final result, i.e. the adult animal, by changing the growth rate of different tissues, or the time of synthesis of certain proteins, speeding up here, slowing down there. Tinkering then operates at the cell level" (Jacob, 1983, p. 141).

CONCLUSIONS

The present situation of evolutionist theories is already different from that in which punctuated equilibria originated in the 1970's as an alternative to classical Neo-Darwinism. At that time a sort of contraposition could be seen between molecular biology, concerned with microscopic evolutionary mechanisms, and paleontology, studying macroevolutionary speciation mechanisms. The important new fact characteristic of the eighties is the overcoming of this contradiction. The search for macroevolutionary mechanisms has met halfway the development of an image of the organism in a more systemic sense. The new theories of evolution are indubitably synthetic, in a more broad and profound sense than the synthesis which they have sought to demolish. Particularly significant are the words with which François Jacob comments on the present state of the theory of evolution, especially if compared with Jacques Monod's *Chance and Necessity*. Whereas Monod considered the mechanisms of evolution as already understood in principle and the problem of evolution as solved, Jacob makes and explicit admission of ignorance. It is, however, a positive admission. The fact that we are faced more with problems to solve than with answers we can already give is a proof of the increase in complexity—and perhaps of validity—in our knowledge.

NOTES

1. Gould (1981) and Prigogine/Stengers (1981) constitute good illustrations of such an interpretation of evolutionary phenomena.
2. Cf. Jacob (1977), (1981).
3. It is precisely on the acknowledgement of the epigenetic and self-transcendent nature of evolutioanry process that are based today some promising attempts to relate biological evolution with evolutionary phenomena of other kinds (cosmologic, thermodynamic, ecological, noologic. . . . Cf. for example Jantsch/Waddington (1976), Jantsch (1980), Jantsch (1981).
4. The dogma of the adaptational origin of the various evolutionary traits is closely accompanied, in many explanation schemes, by the atomistic dogma of the possibility of an "objective" sub-division of the organism into separate traits, independently of any choice by the observer and independently of the total organization of the organism. The question: "what is a trait?" becomes a significant question in contemporary evolutionism. See, on the other hand, the observations of Lewontin cited below.
5. Stephen Gould and Richard Lewontin have sharply criticized many explanations of the socio-biological type encountering in them an acritical adhesion to "dogmas" of this type. Cf. Gould (1977a), Gould/Lewontin (1979), Lewontin (1978), Lewontin (1979).
6. "The adaptationist programme can be traced through common styles of argument. . . . If one adaptive arguments fails, try another. . . . A suite of external structure (horns, antlers, tusks), once viewed as weapons against predators, become symbols of intraspecific competition among males. The askimo face, once depicted as 'cold engineered' becomes an adaptation to generate and withstand large masticatory forces. We do not attack these newer interpretations; they may all be right. We do wonder though whether the failure of one adaptative explanation should always simply inspire a search for another of the same general form, rather than a consideration of alternatives to the proposition that each part is 'for' some specific purpose" (Gould/Lewontin, 1979, p. 586).
7. Cf. Gould/Lewontin (1979), pp. 590–93.
8. This group of mechanisms have a great role in all of evolutionary history. They are mechanisms directly implicated in the evolution of the human species, and mechanisms which consent to *new possibilities* through the narrowness of the *constraints:* "Under the adaptationist program, each major mental attribute is either a direct adaptation for something related (or at least once related to) to survival, or a hypertrophy of the adaptation and, therefore, still based on and constrained by it. . . . But the alternative view that I advocate here, with its focus on *non*adaptation, frees us from the need to interpret all our basic skills as definite adaptations for an explicit purpose. We have inherited, under great constraints to be sure, an ancient structure with some minor modifications. These modifications may have been adaptations for specific functions, but they also engendered a host of non-adaptative consequences. Our large brains, for example, may be minor reconstructions, based on a prolongation of rapid fetal growth rates to later stages of ontogeny. But the magnitude of structural

change may bear little relationships to functional impact, while under the adaptation-ist program we have to view each impact itself as separately selected'' (Gould, 1981, p. 46).

9. We also speak of *chromosomal mutation* to which we may attribute an important role in the processes of quantum speciation. Cf. White (1968), (1978).

10. Stanley (1979), pp. 157–58.

11. Cf. Gould (1977b), pp. 352–409.

12. Cf. the collection of essays and lectures in Waddington (1975).

13. We conclude by attempting to measure the distance which separates the epistemology of Jacques Monod from the affirmations with which François Jacob outlines some future lines of development in the theory of evolution. "To really know how evolution proceeds, it is necessary to understand embryonic development and its limitations. Unfortunately, to this day, very little is known about regulatory circuits and embryonic development. The only logic that biologists really master is one-dimensional. If molecular biology was able to develop rapidly, it was largely because information in biology happens to be determined by linear sequences of building blocks. And so the genetic message, the relations between the primary structures, the logic of heredity, everything turned to be one-dimensional. Yet during the development of an embryo, the world is no longer merely linear. The one-dimensional sequence of bases in the genes determines in some way the production of two-dimensional cell layers that fold in a precise way to produce the three-dimensional tissues and organs that give the organism its shapes and its properties. How all this occurs is still a complete mystery. While the molecular anatomy of a human hand is known in some detail, almost nothing is known about how the organism instructs itself to build that hand, what algorithm it uses to design a finger, what means it finds to sculpt a nail, how many genes are involved and how these genes interact. Even the principles of regulatory circuits involved in embryonic development are not known. Nor is it known whether the arrangement of genes along the chromosomes has any importance for development. As long as the opportunities offered and the constraints imposed by embryonic development are not clarified, it will remain difficult to really understand the rules of evolutionary tinkering" (Jacob, 1983, p. 141).

REFERENCES

BOCCHI/CERUTI [1981], *Disordine e construzione. Un'intepretazione epistemologica dell'opera di Jean Piaget*, Feltrinelli, Milano.

BOCCHI/CERUTI [1985] (a cura di), *La sfida della complessità*, Feltrinelli, Milano.

DARWIN, C. [1872], *On the Origins of Species by Means of Natural Selection, or the Preservation of Favoured Races in the Struggle for Life*, 6th edition, John Murray, London. [Ed. 1952, Encyclopedia Britannica, Chicago].

DOBZHANSKY, T. [1974], *Chance and creativity in evolution*, in DOBZHANSKY T./ AYALA, F.J., *Studies in the philosophy of biology*, Macmillan, London.

GOULD, S. J. [1977a], *Ever since Darwin*, Norton, New York.

GOULD, S. J. [1977b], *Ontogeny and phylogeny*, Harvard University Press, Cambridge (Mass.).

GOULD, S. J. [1981], *On the evolutionary biology of constraints*, in *Daedalus*, Spring 1981, pp. 39–52.

GOULD, S. J. [1983], *Irrelevance, submission and partnership: the changing role of paleontology in Darwin's three centennials, and a modest proposal for macroev-*

olution, in BENDALL, D. S. (ed.), *Evolution from molecules to men*, Cambridge University Press, Cambridge.

GOULD, S. J./LEWONTIN, R. [1979], *The spandrels of San Marco and the Panglossian paradigm: a critique of the adaptationist programme*, in *Proceedings of the Royal Society of London*, B 205, pp. 581–98.

HO, M. W./SAUNDERS, P. T. [1979], *Beyond Neo-Darwinism. An epigenetic approach to evolution*, in *Journal of Theoretical Biology*, 78, pp. 573–91.

JACOB, F. [1977], *Evolution and tinkering*, in *Science*, 196, pp. 1161–66.

JACOB, F. [1981], *Le jeu des possibles*, Fayard, Paris.

JACOB, F. [1983], *Molecular tinkering in evolution*, in BENDALL, D.S. (ed.), *Evolution from molecules to men*, Cambridge University Press, Cambridge.

JANTSCH, E. [1980], *The self-organizing universe*, Pergamon Press, Oxford.

JANTSCH, E. [1981], *The evolutionary vision*, Westview Press, Boulder (Colo.).

JANTSCH, E./WADDINGTON, C. H. [1976], *Evolution and consciousness. Human systems in transition*. Addison-Wesley, Reading (Mass.).

KIMURA, M./OHTA, T. [1973], *Mutation and evolution at the molecular level*, in *Genetics*, 73 (Suppl.), pp. 19–35.

KING, M. C./WILSON, A. C. [1975], *Evolution at two levels in humans and chimpanzees*, in *Science*, 186, pp. 107–16.

LASZLO, E. [1986], *Evoluzione*, Feltrinelli, Milano. English edition, Boston and London: New Science Library, Shambhala, 1987.

LEWONTIN, R. [1977], *Adattamento*, in *Enciclopedia Einaudi*, Vol. I, Torino, pp. 198–214.

LEWONTIN, R. [1978] *Adaptation*, in *Scientific American*, 239, (9), pp. 156–69.

LEWONTIN, R. [1979], *Sociology as an adaptationist program*, in *Behavioral Science*, 24, pp. 5–14.

LEWONTIN, R./LEVINS, R. [1978], *Evoluzione*, in *Enciclopedia Einaudi*, Vol. 5, Torino, pp. 995–1051.

MONOD, J. [1970], *Le hasard et la nécessité*, Le Seuil, Paris.

PRIGOGINE, I./STENGERS, I. [1981], *Vincolo*, in *Enciclopedia Einaudi*, Vol. 14, Torino, pp. 1064–80.

STANLEY, S. M. [1979], *Macroevolution: pattern and process*, Freeman, San Francisco.

STANLEY, S. M. [1981], *The new evolutionary timetable*, Basic Books, New York.

WADDINGTON, C. H. [1975], *The evolution of an evolutionist*, Edinburgh University Press, Edinburgh.

WHITE, M. J. D. [1968], *Models of speciation*, in *Science*, 159, pp. 1065–70.

WHITE, M. J. D. [1978], *Modes of speciation*, Freeman, San Francisco.

CHAPTER 4

Evolution and Intelligence

JONATHAN SCHULL

Editor's Introduction: An additional element in the "changing image" of biological evolution discussed by Bocchi is the enlargement of the concept of the basic units of evolution: in the view championed *inter alia* by Gould and Eldredge it is not only the gene or the individual, but the entire species as well. This, on first sight technical point—which, we should note, is by no means universally agreed among biologists—seems to be of interest only to the life sciences. In fact, it has broad implications that extend to the human sciences as well.

The identification of species as some variety of "individuals" transcends the traditional definition of evolution as concerned uniquely with genotypes and phenotypes. If species (or populations within clades or ecosystems—the point is still in discussion) are themselves units of evolution, the door is opened within biology proper to viewing evolution as a process that concerns evolving *systems*. Such systems may be of various kinds; they need not be limited to genotypes expressed in phenotypes subject to natural selection. The enlargement of the concept of the *unit* of evolution has noteworthy implications for a general *theory* of evolution.

In this chapter Schull, a psychologist versed in biology and in the experimental method, explores the broader concept in reference to the factor of intelligence. He approaches the problem from the perspective of a general theory, avoiding the reductionism that often taints attempts of this kind. Schull suggests a functional model of adaptive behavior that qualifies for the ascription of intelligence, and argues that the individuals that meet its requirements are not limited to humans; they also include biological species. The finding addresses a long-standing problem in the history of biology—that of accounting for the emergence of species through the apparent alternatives of blind chance on the one hand, and purposive planning on the other. Neither alternative is satisfactory, as many biologists well know. It may be that a process common to learning in humans and evolution in the organic world may be at work. If so, evolutionary processes are not merely at the mercy of "blind" mutations, nor are they the manifestations of a higher purpose or teleology.

In seeking to demonstrate that the processes responsible for the emergence of species in biology are isomorphic with those that underlie the processes of learning in humans, Schull prepares the ground for an evolutionary conception that goes beyond biology, without neglecting its current tenets and known processes.

INTRODUCTION

There is an oft-acknowledged analogy between the operation of natural selection in evolving species, and learning through reinforcement in intelligent animals. Other systems (e.g., those responsible for creative human intelligence, human economies, etc.) similarly achieve adaptive self-modification through selective retention or reproduction of successful variants. But how deep is the analogy which relates such systems?

Historically, the question has roots in the work of the pre-Darwinian biologist William Paley (1802), who argued that the adaptedness of living things proves the existence of an intelligent designer, namely God.

> There cannot be design without designer; contrivance without contriver; order without choice . . . relation to a purpose without that which could intend a purpose. . . . Arrangement, disposition of parts, subserviency of means to an end, relation of instruments to a use, imply the presence of intelligence and mind. (Paley, 1839, p. 10).

Ever since Darwin showed that purely materialistic processes could account for adaptedness, Paley's theological argument has been dismissed. But it has not been refuted: materialism is no longer deemed incompatible with intelligence and mentality, so why do we not entertain the notion that intelligent design is indeed responsible for the exquisite creations of the biological world? (cf. Bateson, 1972, 1979.)

Recent developments in biology, philosophy and psychology likewise suggest that Paley's argument warrants secular reappraisal. Contemporary biologists and philosophers of biology now claim that species are best seen as individuals (rather than natural kinds) localized in a particular spatio-temporal domain, and unified by a highly organized yet continually changing gene pool (Ghiselin, 1974, Hull, 1974). But in light of the analogy between evolution and learning, this claim raises an unanticipated possibility: species, like animals and humans, might usefully be seen as intelligent (but not omniscient), purposeful (but within evolving limits), and cognitive (but with ample capacity for error). This possibility is only strengthened by the growing recognition of the importance of hierarchically organized variation/selection systems in both biological evolution and human creativity (see, e.g., Dennett, 1975, and Simon 1962).

This paper will attempt to examine this issue and critique the conventional theoretical frameworks which make the possibility of intelligent species seem outlandish (c.f. Schull, 1990, which deals in more detail with theory *per se*.) It will not, however, attempt to deal with the much more obscure issues of consciousness or awareness (c.f., Laszlo, 1981).

Nothing which follows should be construed as arguing that species (or for that matter, animals) are sentient. Rather, I want to discuss interactions and relations among two biological processes which are directly or indirectly responsible for the existence of intelligent agents on earth—learning and evolution—and show how these interactions and relations provide an explanation for the "character" and capabilities of intelligent animals, and of the evolving species they comprise. For reasons that will hopefully become clear, I will use the term "intelligence" to refer to the ability of a system to devise new adaptations to novel circumstances. I will use the term "mental processes" to mean those processes which mediate intelligence. My aim is to clarify the nature of those processes and to examine the possibility that they play a role in species evolution as well as in animal and human learning.

INTELLIGENCE

We can begin as William James began his *Principles of Psychology*, by pointing to the phenomenon we seek to explain: purposive intelligence and the existence of organs, organisms, and/or actions which seem to be designed for, or motivated by, purposes. James observed that

> with intelligent agents, altering the conditions changes the activity displayed, but not the end reached. . . . The pursuance of future ends, and the choice of means for their attainment are thus the mark and criterion of the presence of mentality in a phenomenon. (James, 1890, p. 5)

Today, we know a great deal about the kinds of systems which display such performances. We are also a great deal more confused about the kinds of systems which warrant the attribution of purposive intelligence, and/or mentality (see, e.g., Dennett, 1983, Searle, 1980, and Block, 1981). But James' own examples will be useful and, hopefully, non-controversial.

> Blow bubbles through a tube into the bottom of a pail of water, they will rise to the surface and mingle with the air. But if you invert a jar full of water over the pail, they will rise and remain lodged beneath its bottom, shut in from the outer air, although a slight deflection from their course at the outset, or a re-descent towards the rim of the jar when they found their upward course impeded, would easily have set them free. (p. 4)

In contrast, suggests James, suppose that a living frog be trapped underwater in a similar situation.

> The want of breath will soon make him . . . take the shortest path to his end by swimming straight upwards. But if a jar full of water be inverted over him, he will not, like the

bubbles, perpetually press his nose against its unyielding roof, but will restlessly explore the neighborhood until by re-descending again he has discovered a path around its brim to the goal of his desires. Again the fixed end, the varying means! (p. 4)

Now, regardless of how much intelligence we wish to attribute to a lowly frog, the distinction between the frog and the bubbles is important. If we were to build a cybernetic automaton with the capabilities of James' frog, there should be little dissent to the proposition that such an automaton would be, in some sense, more intelligent than one which merely mimicked the behavior of an air bubble, and less intelligent than a system which could additionally learn new ways of achieving its goal.

The questions at hand, then, are (1) how have processes of evolution and learning given rise to various grades of intelligence such as these, and (2) what grade of intelligence might an evolving species possess?

A MODEL

I will begin by describing a proto-purposive system which neither learns nor evolves but which is a model of learning (the acquisition of new adaptation by an individual animal during its own lifetime) and of evolution by natural selection (the acquisition of new adaptations by a species during the course of many generations of individuals). The mechanisms of learning (nervous systems and developmental environments) and those of evolution (genes and much, much more) are assuredly very different. But their dissimilarities should not blind us to more fundamental similarities in the kinds of processes which make adaptive systems what they are.

The proto-purposive system is a hypothetical electronic "bug", complete with legs and two forward-looking eyes, hard-wired such that it always turns toward whichever of its eyes is more brightly illuminated. Construct such a bug from photocells and electric motors, and it will be positively phototropic: put it down on the floor of an empty room with a single bright light on one wall, and the creature will turn toward the light and move unerringly to it. Displace it in space or orientation and it will compensate, adopting a new course to its fixed "goal."

The bug can also stand for the infinitely more complex phenomena of learning through reinforcement and evolution through natural selection. In both cases, the essence of the phenomenon is that the system is sensitive to contingencies of selection, i.e., the effects of the environment upon the adaptive system after its variant response. In learning, these are known as contingencies of reinforcement; in evolution they are known as selection pressures. Let us adopt the language of learning first.

In the study of learning, positive reinforcement is said to occur when a response produces an environmental consequence which in turn strengthens the response itself; punishment is said to occur when a response produces a consequence which weakens the response.

Now, here is the sense in which the bug instantiates these contingencies: Place the bug on the floor, facing exactly away from the light. In this situation, the illumination of both eyes is equated and so the bug blithely marches away from the light until some deviation in its course occurs. But suppose some variation in its physiology or its environment causes the bug to deviate briefly rightwards (the response). Relative to the bug, this shifts the light to the side such that it now casts more light on the bug's right eye (the environmental consequence). This will reinforce the act of turning rightward, until, facing the light, the bug now moves straight ahead. This is analogous to positive reinforcement in the sense that the environment's reaction to the animal's response of turning toward the light causes the response to be strengthened and continued.

The other contingency immanent in this model is "punishment." Suppose that while heading toward the light, the bug again veers right. In this situation the light moves from straight ahead, to a position leftward of the bug. The left eye is more brightly illuminated and the response is "punished": the light causes the bug to undo its response, and to turn left, back toward the light. As in animal learning, then, the system is built in such a way that certain responses (movements toward the light) are strengthened, while other responses (movements away from the light) are weakened.

Obviously, because these contingencies have no long-lasting effects upon the organization of the bug, the whole situation is a caricature of what goes on in learning. But the model makes plain what can otherwise be obscured: behavior and adaptiveness is a property of the interaction of organism and environment. In any goal-oriented system, moment to moment behavior is determined by relationships between the organism and its environment; knowledge of both of these components is necessary for the prediction of the system's behavior; and environmental organization no less than organismic organization co-determines the bug's "behavior." The phototropism is a property not of the bug, nor of the environment, but of the bug-in-the environment (see Ashby, 1960 for a brilliant and extensive exposition of this issue).

The conclusion is quite general. While real organisms are considerably more complex than the bug (and therefore their internal organization may be relatively more important than the environment's) the physical environment in which functional behavior occurs is part of the physical substrate or "material mechanism" of that behavior. Changing either component could change the animal's molar behavior in the same way: e.g., you could make the animal reverse its tropism either by crossing the wires connecting each eye to the steering apparatus or by altering the surface upon which the bug moved such that motor movements that previously caused right turns would now produce leftward locomotion, etc. And changing both of these components in concert would leave the bug's behavior intact, thus demonstrating that it is not organismic or environmental structure per se which determines behavior, but rather the relation between them.

My claim is that it is the kinds of interactions and relations which go on between an adaptive system and its environment which are criteria in characterizing it as intelligent or not, rather than its physiological makeup. To the extent that an evolving system and an animal of a certain degree of purposive intelligence have the same kinds of interactions with, and responses to, their environment, to that extent are they of comparable purposive intelligence.

The problem is to find a way of comparing the "behavior" of animals and species, but that, too, is the function of our "bug model." Move this model up one level of abstraction and you have a description (but not an explanation) of evolution by natural selection. Let the two dimensional surface on which the bug travels stand for a multi-dimensional "trait space." Then, the characteristics of a population could be located somewhere on this surface. The "bug" can now stand for a population of organisms occupying some particular place on this surface (i.e., with a particular set of characteristics) and the light can stand for maximal fitness. Indeed, if we move the light far overhead, put hills in the trait surface, and let the vertical dimension represent the fitness of each point on the surface, then we have recreated the familiar metaphor of the adaptive landscape (see, e.g., Eldredge, 1985) and our bug will behave the way adapting species are said to behave, moving imperturbably toward the light, and finding its way to adaptive peaks—to local fitness maxima. As in the model, any mutations (i.e., displacements in the multidimensional "trait-space") which constitute deviations away from fitness will be counteracted, while others which move the animal toward maximum fitness will be amplified. Just as the bug tends always to move in the direction of the light, so, according to the theory of natural selection, does a reproducing population move inexorably through a multidimensional trait space in the direction of increased fitness.

THE THEORY OF NATURAL SELECTION

There has been much debate about the extent to which the theory of natural selection explains anything at all, and if so, what. I believe our model allows us to see clearly what kind of explanation it provides. Natural selection explains exactly as much about the behavior of an evolving species as a wiring diagram explains about the behavior of the bug. At the same time, the theory says very little about the actual evolutionary trajectory of an evolving species; as with the bug, that is determined by the ongoing particulars of the system's interactions with its environment.

Darwin's (1859) great achievement was to show that a mechanism capable of producing adaptive, seemingly purposeful evolutionary change was immanent in certain undeniable facts of life. In essence, the argument is simple. For a given species in a given environment, the number of individuals tends to remain constant (Fact 1). And yet, in each generation there tend to be more offspring than parents (Fact 2). Therefore in each generation some offspring die without reproducing, while others live to become parents. There is, in short, competition among individuals in populations, and losers do not leave offspring (Deduction 1). Finally, individuals vary, and offspring tend to resemble their parents (Fact 3), and so in each generation the offsprings will tend to resemble the more successful parents of the previous generation (where "success" or "fitness" means reproductive success) (Deduction 2). This is natural selection, and it means that for any species which exists for a long enough time in a given environment, characteristics associated with reproductively successful individuals will persist and accumulate. The Earth will inherit the fit, because natural selection, a process entailed by the three facts about organic populations reproducing in natural environments, "acts constantly to improve the adjustment of animals and plants to their surroundings, and their ways of life." (Huxley, 1942, reprinted in Appleman, 1970, p. 326)

The theory of natural selection explains how, by subtle but purely physical means, the structure of the environment can constrain inheritable characteristics in populations of organisms. The three facts are analogous to our description of the bug's hard-wiring, and the two deductions correspond to a proof that the behavior of any bug so constructed will be systematically constrained by the structure of the environment to find its way to the top of the nearest hill.

The phenomenon of fitness that Paley and Darwin found so telling—
the implausible occupation of adaptive peaks by biological organisms—

is thus explained. But it is not explained away, nor is it rendered unimpressive. The trait space through which a species moves is far more complex than that through which the bug moves, and the repertoire of responses made by the species is far richer.

The question remains: how intelligent and purposive is the behavior of an evolving species, as compared to the bug, and to animal learners? To answer the question we need to examine the characteristics of animal learners and of the learning processes which generate various grades of intelligence. We have two tasks: to review the standard version of selectionist theory which has been developed to explain the acquisition of behavioral adaptations, and to confront the claim, made by some behavioristic developers of that theory, that their approach eliminates the need for mentalistic notions of intelligence and purpose in understanding behavior. (In fact, behaviorism has fallen on hard times, and for reasons which suggest that mentalistic notions could be useful in understanding evolution as well as the problem of psychology.)

THE THEORY OF LEARNING BY REINFORCEMENT AND ITS LIMITATIONS

B.F. Skinner, the last great proponent of behaviorism, has explicitly acknowledged that his use of the concept of reinforcement to exclude mentalism from psychology is closely parallel to Darwin's use of selection to argue against Paleyan teleology in Biology. He notes that the effect of contingencies on the probability of a given response in a given situation

is due, in phylogenic behavior to the selection of genotypes, and in ontogenic behavior, to operant conditioning. The tendency to behave in a given way on a given occasion has been attributed to instinct in the phylogenic case and habit in the ontogenic case. In both, it has been associated with the concept of purpose, and in ontogenic behavior with expectation or intention. Concepts of this sort add nothing to the observed facts, and they cause trouble, because, by referring to inner determiners of behavior they often serve as substitutes for the further explanation that will eventually be provided by physiology. (Skinner, 1975, p. 120)

But this claim loses its force if physiology is only part of the story, and if the "further explanation" necessarily involves processes which play an explanatory role which can not be played by physiological or mechanistic processes alone. I have already argued that the environment as well as physiology plays an active role in guiding purposive behavior; I will now elaborate with regard to learning in particular.

Thorndike's original account of the acquisition of adaptive behaviors was explicitly modeled after Darwin's theory (Thorndike, 1898). He supposed that specific responses were generated by stimulus-response (S-R) bonds. Random shifts in the strength of the various S-R bonds, and changes in environmental stimulation (some of them random, some a result of the animal's behavior) would produce changes in the probability of various responses. However, certain behavioral consequences (those produced by S-R bonds which give rise to ineffectual or maladaptive behaviors) would become relatively inactive. An animal in a given situation would gradually engage in behaviors which result in rewards, and would gradually abandon behaviors which had been ineffectual (or worse). Thus, the seemingly purposive behavior of an animal "working for" future rewards was reinterpreted as the strictly mechanical result of past rewards which preserved some S-R bonds at the expense of others. Now, we are still fairly ignorant about the physiological nature of the reinforcement process itself (as compared to our substantial post-Mendelian understanding of the analogous population-genetic processes of natural selection), but there is little doubt that some such process could be instantiated in a nervous system. And so, as with Darwin's 'refutation' of Paley, behaviorists use reinforcement theory to argue against the use of non-physiological (i.e., mental) explanations of behavior.

However, as Skinner himself emphasized, Thorndike's account rests on the fiction that the only learned responses which animals make are ones that have been previously reinforced. In fact, instrumental behaviors are quite variable: the actual muscle movements by which the animal obtains reinforcement vary from trial to trial, and new variants crop up all the time. Skinner's ingenious solution to this "problem of the generic response" was to redefine the phenomenon to be explained. The "operant response" in Skinner's terms is actually part of the environment, specifically the consequence of the behavior (e.g., the depression of a lever) which procures reinforcement. It is operants, then, which are reinforced, which are of fixed topography, and which obey regular laws of behavior. Skinner notes that the "obvious generic nature of the response as measured behaviorally raises an acute problem for neurology which for the most part may be ignored in behavior." (1938, p. 430). But he is wrong in implying that he has solved, even provisionally, the problem of explaining how organisms manage to behave the way they do. Rather, he has relegated to others the fundamental problem of explaining how organisms manage to generate the actual behaviors (left-handed bar presses, right-handed bar presses . . .) all of which converge somehow on the environmental contingencies. However, this is none other than the problem of explaining how animals choose varying means for the attainment of future ends (James' "mark and criterion of the presence of mentality in a phenomenon,") and until that problem is solved, Skinner's assertion that "behavior which seems to be the product of mental activity can be explained in other ways," remains an unsupported article of faith.

Moreover, while Skinner hopes to hand his "acute problem" to the neurologist, we have already seen that wiring diagrams per se do not explain actual behavioral trajectories, while interactions-with-environments do. Behavior (and learning) *is* the interaction of a nervous system with its environment. It is these organism-environment interactions which underlie and embody the purposive intelligence of individual organisms. And, as we shall see, they may provide the solution to the generic response problem while deserving to be called "mental processes."

EXTENDING SELECTIONISM IN LEARNING

The solution to our problem is implicit in James' principles of psychology, and is explicit in an article on the law of effect by D. C. Dennett (1975). Suppose, Dennett writes, "that creatures have two environments, the outer environment in which they live, and an 'inner environment' which they carry around with them. The inner environment is just to be conceived as an input-output box for providing feedback for events in the brain" (p. 77) and its function is analogous to that of the physical environment which provides feedback for behaviors actually emitted by the animal.

In other words, before a rat reaches out to press a bar, it generates many potential plans of action, and then selects a motor plan which meets certain cognitive criteria for probable success in the current situation. More generally, the suggestion is that the inner environment selects, from all the potential plans-of-action which are generated, those "good bets' which meet its criteria for selection. Such a mechanism provides a solution to the generic response problem and explains how "we, and even monkeys often think out and select an adaptive course of action without benefit of prior external feedback and reinforcement." (Dennett, 1975, p. 85; c.f., Campbell, 1974) The solution is non-behavioristic, since most of the important selection does not involve motor behavior. But it is a perfectly plausible and physically realizable natural-selection-like process.

In its most parsimonious form, the theory supposes that intelligence of this sort (the ability to produce new adaptive behaviors without having to test them in practice first) is the result of a trial-and-error simulator which involves the selective "ratifying" of behavior plans "nominated" by the random activation of available neutral elements. Only those plans which happen to meet the operative criteria for selection ("will they pay off?") are put into action, and the result is a seemingly insightful creature. In an elaborated (and for more intelligent organisms, more realistic) form, the theory further supposes that the selector and the generator-of-variants are themselves products of variation and selection processes which have operated in the past within the organism, between organism and environment, or between species and environment (i.e., which are innate). The important point is that successive layers of variation and selection provide the system with greater degrees of adaptability and give it the ability to produce more impressive displays of learning, learning-to-learn, intelligence, and insight.

Because this is an idea which we will be exploiting forthwith, two points should now be understood. First, the theory is not new. James supposed that the adaptive function of consciousness was to "reinforce favourable possibilities [for action] and repress the unfavourable or indifferent ones" (p. 93). Herbert Simon (1970) suggested that all human problem-solving is due to nested hierarchies of variation and selection. Indeed, Dennett (1978, p. 82) and Bateson (1979) have suggested that something of this sort is necessarily the only well from which intelligence and creativity can be drawn. Second, what is "inner" about the inner environment is not its location within the animal's nervous system (for it is not in fact entirely intra-neural), nor any supposed independence from the physical environment (for it is a product of the evolutionary and/or developmental environment). Rather, the inner environment is "inner" in the sense that it is nested within a larger variation-selective system, involving both developmental selection (learning) and evolutionary selection (à la Darwin).

GRADES OF INTELLIGENCE

Now let us try to map the various degrees of intelligence under discussion onto the behavior of a creature like our bug. The original bug did not in fact learn, manifesting only a simple tropism. Like the air bubble trapped under the jar, it would be completely stymied if a barrier were placed across its path toward the light. (Let us suppose that the barrier is a fence high enough to impede forward progress, but not so high as to block the bug's "view" of the light.)

What would it take to build more intelligence into the bug? Variability of response, for one thing. If thwarted progress increased variability, then the bug, like James' frog, might accidentally escape around the edge of the fence if it emitted a random turn to the left. But if it came up against another fence, its behavior would not show improvement. The bug cannot learn.

In contrast, a more intelligent bug would be equipped with the ability to discriminate more than one stimulus (e.g., the fence as well as the light), and would have the ability to be reinforced (à la Thorndike). Then, its behavior would be different; after accidentally escaping around the left edge of a fence on one occasion, the bug's behavior would be modified such that the next time it came up against a similar impediment it would turn left with a shorter latency than before.

That would be smart, but not overwhelmingly so. Because it lacks a solution to the generic response problem, a Thorndikean bug would still be so stupid as to turn left (as it has been consistently rewarded for doing), even if it were much closer to the right edge of the fence. (A chronically variable Throndikean bug might well learn to turn right when near the right edge, and left when near the left edge, but it would have had to stumble onto those responses and be rewarded for them in those settings.) It would also have to blindly generate large numbers of random (and often maladaptive) responses before it happened to hit upon an adaptive one.

In contrast, a Dennettian bug further equipped with a generator-of-potential-responses, as well as an inner environment which could simulate the essentials of the situation, would, having appraised the distance to the two ends of the fence, simulate the consequences of a leftward turn, simulate the consequences of a rightward turn, select the plan of action which would be reinforcing, and then (finally) do the right thing, without having had to discover it empirically through tedious trials and expensive errors. Such performances are often called "insightful."

To extend this approach one more time, let us admit that the last creature, while avoiding the costs of trial and error, would spend an inordinate amount of time "lost in though," if it had to generate and reject all possible responses (but one) at every choice point. With repeated exposure to a recurring situation, an even more intelligent animal would select among a very large set of possible-response-generation-schemes a much smaller number which would serve to generate responses needed in that situation. Mechanistically, this is just one more level of variation and selection: the operation of one inner environment is now modified by selection at another level. Behaviorally, such an animal could be recognized by its ability to progress, on repeated exposure to a novel situation, from insightful solutions which took some time to achieve, to a more routinized but shorter latency adoption of stock solutions to now-standard problems.

To turn the argument around (and this is the point of this exercise), an adaptive system which was able to demonstrate objective behavioral performance like the one just described would be more intelligent than a "merely" insightful system, which would in turn be more intelligent than a Thorndikean creature, which is in turn more intelligent than our original purposive, but severely limited, bug.

Note that at each iteration we need to add to the system two things: an additional layer of variation-selection, and an increased capacity to process information about the situation. The information processing requirement is an obvious requirement for any system deserving the attribution "intelligent," but it is by no means a sufficient one—a claim supported by critics of AI who ridicule computer systems which process vast amounts of information but do so in a rotely mechanical fashion (Searle, 1980; Block, 1981). Multiple layers of variation-selection, however, are a different matter. They, along with the requisite capacity to process information, may well be both necessary and sufficient for the attribution of intelligence. Evolutionary epistemologists like Donald Campbell (1966) have long argued that all new knowledge must come from variation and selection. It is interesting to note that systems which generate intelligent behavior by such means are immune to critics of "brute force" artificial intelligence who charge that intelligence is embodied in rules developed by human programmers, and not by machines.

On the other hand, in contrast to one such critic (Searle, 1980) who resorts to the speculation that nervous systems are uniquely able to mediate "real" intelligence, I wish to advance the thesis that it is not the kinds of tissues which matter, but rather the kinds of organism-environment interactions which distinguish real intelligence from simulated intelligence. The kinds which are crucial are the kind we have been discussing, whether made possible by neurons, silicon chips, or populations of organisms. By this criterion, brute force computer simulations do not qualify as truly intelligent (which is intuitively right), but programs like the General Problem Solver (Newell and Simon, 1963) might. Humans and many animals do qualify. And evolving species may qualify as well.

EXTENDING SELECTIONISM IN EVOLUTION

That species process vast amounts of information is indisputable. One need only consider the manifest fit of organisms to their niches, the mastery of optical principles demonstrated by the design of the eye, and the existence of insects whose bodies are exquisite facsimiles of the leaves on which they forage. The pattern of leaf-like veination on such an insect, for example, was transmitted to the insect's genome through selection pressures exerted by the insect's predators, and was built into the predators' nervous systems by selection pressures exerted by the exigencies of foraging for camouflaged prey. Yet the pattern is reproduced with spectacularly high fidelity. As processors of information relevant to reproductive success, then, populations-in-environments are at least comparable to any known information processing system, animal or artificial.

But do they show multilevel adaptive capabilities comparable to the more complex creatures described above? There seems to be a problem. We have characterized our original bug as a non-learning creature, grossly inferior to other, more elaborate and intelligent systems. And yet the adaptive-landscape metaphor also likens an evolving species to our original bug. How then, can we conceive that an evolving species may be more than minimally intelligent? The answer is that the adaptive landscape metaphor is seriously misleading. To demonstrate this, I will adopt the analogous representation to describe the adaptive changes undergone by various grades of our bug. We will see that the systems we just examined, however sophisticated, all look the same in this representation, an appearance which obscures the most important differences among them.

The adaptive landscape metaphor in biology is one in which each dimension (except for the fitness dimension) represents the genetic composition of the species. For a learning organism, the analogous ontogenetic adaptive landscape would stand for the organization of the animal's nervous system (analogous to the species' gene pool). Because the nervous system of a hard-wired bug does not change, the point representing such a bug on an ontogenetic adaptive landscape would not move. In contrast, the point representing adaptation by a Throndikean bug *would* move up hill, as the animal increased its rate of reinforcement through adaptive change in its nervous system; so would the point representing adaptation by a Dennetian creature with an inner environment, and so would the point representing adaptation by even more sophisticated creatures. All adapting systems look the same, no matter how different their means of adaptation. Thus, while the notion of the adaptive landscape is a useful way of describing evolutionary trajectories, it systematically obscures the character of the process of adaptation (much as the population genetic approach does in general).

Nonetheless, the model makes it clear that an evolving species is, at the least, the equivalent of a Thorndikean animal. It is certainly as resourceful as the Jamesian frog: an evolving species confronted with an ecological circumstance to which it is not adapted (analogous to the fence blocking our bugs' progress) will emit novel responses, thus increasing the probability of a new adaptation. This increase in variability will occur for several reasons. First, developmental homeostasis will falter in the absence of its usual environmental supports (Ho and Saunders, 1979; c.f., Ashby, 1960). Second, for similar reasons, genetic mutation rates may or may not increase. And third, in species which usually reproduce asexually, genetic recombination rates increase precipitously, because these are the circumstances under which sexual reproduction becomes the predominant strategy (Bell, 1982). Working with the variants thus generated, the species will act like a Thorndikean creature which learns to repeat variants if they are successful: if any of the novel responses happen to provide a way of circumventing the barrier to fitness, they will be reinforced (through differential reproduction). And, like a Thorndikean animal which differentially emits certain responses in the circumstances in which they have been rewarded, the species will manifest the new responses in the new circumstance more readily than in the earlier one (because environmental stimuli which are present in the new circumstance, and which are possibly necessary for the expression of the new adaptation, are lacking in the earlier situation). In learning, the phenomenon is known as conditional discrimination; in ecology, it might be called an "ecologically determined polymorphism", or a kind of mixed strategy.

In terms of intelligence, this is not negligible, yet by this account (which is orthodox neo-Darwinism) the species would still have the same limitations as the Thorndikean animal. It must generate a large, undirected, and usually maladaptive set of random genetic mutations before it can hit, purely by chance, on an adaptive one. Moreover, fine tuning of the adaptive response to specific circumstances (analogous to the rat's ability to reach for the bar in a manner appropriate to its particular orientation in the experimental chamber) must depend upon the same process of blind trial and frequent error.

This undirectedness is taken by many (e.g., Gould, 1982a; Monod, 1970) to be a key feature of the classical Darwinist portrayal of the evolutionary process. But the fact is that while undirectedness is characteristic of genetic mutations, it is not characteristic of the generation of adaptive responses, nor of evolutionary changes within the gene pool. The generation of adaptive responses, and the trajectory of phenotypic changes in the population is guided, not just by species-independent selection pressures, but by the species itself. Just like the Dennetian animal which can screen potential responses "internally" by filtering them through an "inner" environment, every species contains a host of subsystems which generate potential adaptive responses and evaluate them in a preliminary fashion before committing the species to genetic change. These subsystems are individual organisms, and the screening process is ontogenetic adaptation.

PHYLOGENY RECAPITULATES ONTOGENY

Because ontogenetic adaptation is so much more rapid than evolutionary adaptation, it will necessarily be the vanguard of a species' response to novel circumstances. Individual animals, born into an environment to which their congenital adaptations are only marginally adequate, will cope as best they can. A variety of new, potential adaptations will be generated even in the absence of genetic variation for those adaptations, because no organism, plant or animal is entirely without the capacity to adapt to local circumstances. However, when those circumstances are not the ones for which the organism's genes have been naturally selected, the responses which emerge will be unusually variable and unpredictable—due to chance, to the aforementioned fluctuations in non-canalized epigenetic pathways, and to variations in local circumstances.

These individual developmental responses, even if random and undirected (which they are not), will necessarily lead to directed and non-random evolutionary change, for reasons which have been argued by Baldwin (1896), Bateson (1963), Braestrup (1971), Ho and Saunders (1979), Piaget (1978), and Waddington (1975), among others. The phenomena in question are variously known as phenocopy, genetic assimilation, and the Baldwin effect. The argument goes like this: with time, and perhaps within a single generation, the population will settle upon those few developmental responses which serve best in the situation. This winnowing of developmental responses need not involve natural selection of genetic variants; it could occur because each individual organism eventually arises at the best available or most readily achievable combination of physiological parameters available to it, or because individuals perish if they fail to hit upon a viable mode of adaptation soon enough. In any case, a population's response to a novel ecological situation will often be achieved first in the absence of a differential survival of genes which predispose it to the adaptive response. But population-genetic change will follow over a much longer time period, for the simple reason that the new adaptive strategy will automatically create a new array of selection pressures which, over a much longer time period, guide subsequent population-genetic change in the direction of genotypes which make that developmental adaptation easier, faster, and less costly to achieve. The result would be, in Bateson's (1963) term, the genetic "ratification" of an adaptation which was already developed and tested, through non-genetic means. Like the Dennetian organism which avoids tedious trial and expensive error through pre-evaluation of potential responses, the gene pool makes its move only when a viable solution has already been found.

Additional hierarchical levels of variation and selection are of course conceivable, and in all probability of real importance (though they are woefully unstudied). It is being increasingly recognized that selection pressures, organisms, and genomes are all hierarchically organized within monospecific populations (Reidl, 1979; Gould, 1982b). Species are comprised of subspecies, races, local populations, clans, families, and individuals; and within the individual organism adaptive mechanisms are hierarchically organized right down to the level of RNA (Weiss, 1970; Rozin and Schull, 1988). Since we are primarily concerned with relations between learning and evolution, I will focus on the phylogenetic consequences of learning, but it should be remembered that learning is only one of the important modes of ontogenetic adaptation, although in some species it is certainly of key importance. Modern biology's rejection of anything even reminiscent of Lamarckism has apparently led to a serious neglect of process originally envisioned by Baldwin and Morgan (see Braestrup, 1971). A simple example involving instrumental behavior is describing by Skinner (1969).

Behavior arising from ontogenic contingencies may make phylogenic contingencies more or less effective. Ontogenic behavior may permit a species to maintain itself in a given environment for a long time and thus make it possible for phylogenic contingencies to operate. There is, however a more direct contribution. If, through evolutionary selection, a given response becomes easier and easier to condition as an operant, then some phylogenic behavior may have had an ontogenic origin. One of Darwin's "serviceable associated habits" will serve as an example. Let us assume that a dog possesses no instinctive tendency to turn around as it lies down but that lying down in this way is reinforced as an operant by the production of a more comfortable bed. If there are no phylogenic advantages, presumably the readiness with which the response is learned will not be changed by selection. Buy phylogenic advantages can be imagined: such a bed may be freer of vermin, offer improved visibility with respect to predators or prey, permit quick movement in an emergency, and so on. Dogs in which the response was most readily conditioned must have been most likely to survive and breed. (These and other advantages would increase the dog's susceptibility to operant reinforcement in general, but we are here considering the possibility that a particular response becomes more likely to be conditioned.) Turning around when lying down may have become so readily available as an operant that it eventually appeared without reinforcement. It was then "instinctive." Ontogenic contingencies were responsible for the topography of an inherited response (p. 203–4).

Thus in evolution the adapting population of animals plays a role analogous to that of the cortex in individual mammalian behavior; interacting with the environment to develop new patterns and try them out, and then passing control of routinized patterns to more automatic and less plastic centers of interaction with the environment (subcortical systems in individuals, genetic "programs" in species). This is comparable to the highest-grade bug we considered earlier.

Further elaboration is certainly possible: after all, as we have seen, learning is itself hierarchically organized in nested levels of variation and selection.

But perhaps the essential point has been made: the standard neo-Darwinian representation of anagenesis systematically obscures the sophistication and complexity of the processes by which species evolve. A species is every bit as complex as, say, a monkey, and judging by the way it works (i.e., the design of adapted and adapting physiological individuals) it is not wanting in practical intelligence. Moreover, because the ontogenetic adaptations achieved by individuals must often function to guide the evolution of the species, the intelligence of individuals are part of, and not just the products of, the resources by which a species adapts.

That a learned behavior can create selection pressures leading to the genetic "institutionalization" of the same behavior is interesting for another, more subtle reason as well. It shows how the behavior, a dynamic pattern of interaction between organism and environment, is more stable than, and also determinative of, the more "solid" material mechanisms which make it possible. The mechanisms change during the evolutionary scenario, from ontogenetic adaptation to innate adaptedness. Meanwhile the behavior itself, while guiding these changes, remains essentially unchanged. There is nothing mysterious about this, but the example illuminates an analogous phenomenon which does seem quite mysterious: the phenomenon of mind.

The relation of thoughts to behaviors is similar to the relation of ontogenetic adaptation by a population to evolutionary change by it. Thoughts, ideas, intentions, etc., are much more evanescent than their relatively solid physical mechanisms and manifestations. Yet thoughts somehow both depend upon, and influence, their physical correlates. This analogy is worth noting, for it suggests that the ascription of intelligence to an organic species may be more than simply a way of calling attention to the richness of adaptive processes.

SHOULD SUCH PROCESSES BE CALLED "MENTAL"?

The grounds for applying the phrase "mental processes" to such inter-actions between individuals (animal or species) are several.

First, since mental phenomena such as beliefs, preferences, attitudes, knowledge, etc. are fundamentally "about" relations of an organism and its behavior to its environment, and since those phenomena have physical counterparts in organism-environment relationships and interactions occurring over many time scales, it seems to me that prospects of re-ducing "mind" to matter are improved if the material processes in question are not restricted to neural processes in a part of an organ-ism. Second, these processes provide a non-behavioristic account of intelligent, purposive behavior. And third, they provide a rationale for several intuitions which are widely held with regard to mental processes, but which are otherwise unsupportable from a scientific point-of-view.

The first intuition is that mind is not the same as brain. In this regard, I have tried to show that the processes which produce adaptation impor-tantly involve the environment. Selectionist theories necessarily invoke environmental patterns of organization which directly or indirectly con-strain the activity of organismic mechanisms. It follow from this that the "machine" which produces behavior is more than the physiological or-ganism itself, and that much of the machinery lies outside the organism's skin (Ashby, 1960). To put it another way, the brain's structure is such that much of the organized constraint actively determining behavior arises through past and present relationships and interactions with the environment. These relations and interactions are every bit as real as the strictly intra-neural interactions which make them possible, but they are not identical to, nor reducible to, those neural mechanisms.

The second intuition is that mind can influence brain. I have tried to show that mental processes (construed as organism-environment interac-tions with selective consequences) can in fact influence and guide mate-rial neural processes which are more transient, more localized, and less reliably associated with intelligence than the patterns of interaction they make possible. (Just as a species' ontogenetic actions guide and influ-ence its phylogenetic evolution.)

And the third intuition, held by most cultures and at most times, is that mental processes have played a role in the design of the biological world. If mental processes are construed in the way I am suggesting (cf. Laszlo, 1981; Valsiner, 1984), and if the intelligence of an evolving species is indeed at least comparable to that of a highly intelligent animal (as I have argued), then there is a sense in which this last intuition is importantly true.

In principle, then, I see no *a priori* grounds for rejecting the scientific hypothesis that intelligence and mind are involved in biological evolution. The empirical evidence is no less suggestive now than it was in Paley's time. The intelligence in question is not timeless, transcendent, and perfect, as envisioned prior to Darwin, and the nature of the supposed mentality is quite unclear, but the matter is worthy of a good deal more attention than it currently gets from philosophers, psychologists, biologists, and cognitive scientists.

If nothing else, it should be recognized that the behavior of the individuals called species is far more sophisticated than the neo-Darwinian portrayal seems to imply (cf., Campbell, 1982). The cognitive revolution in psychology has revealed new vistas in the study of animal and human adaptability. A similar revolution in the way we view the processes responsible for evolutionary change may be required before we can fully appreciate the workings of evolution.

REFERENCES

Ashby, W. R. (1960) *Design for a Brain*. London: Chapman and Hall.

Baldwin, J. M. (1896) A new factor in evolution. *American Naturalist* 30: 441–451. 536–553.

Bateson, G. (1963) The role of somatic change in evolution. Evolution 17:529–539. Reprinted in Bateson, G. *Stepts to an ecology of mind*. New York: Ballantine, 1972.

Bateson, G. (1972) *Steps to an ecology of mind*. New York: Ballantine.

Bateson, G. (1979) *Mind and nature: a necessary unity*. New York: Bantam.

Bell, G. (1982) *The masterpiece of nature: the evolution and genetics of sexuality*. Berkeley: University of California Press.

Block, N. (1981) Psychologism and Behaviorism. *Philosophical Review* 90:5–43.

Braestrup, F. W. (1971) The evolutionary significance of learning. *Vidensk, Meddr dansk naturh. Foren.* 134:89–102.

Campbell, D. T. (1966) Evolutionary epistomology. In: P. A. Schilpp (Ed.) *The philosophy of Karl R. Popper*. The Open Court Publishing Company (The Library of Living Philosophers), La Salle, Illinois.

Campbell, J. H. Autonomy in evolution. R. Milkman (Ed.) *Perspectives on evolution*. Sunderland, MA: Sinauer Associates.

Darwin, C. (1859) *On the Origin of Species*. London: J. Murray; Harvard: Harvard University Press, 1964.

Dennett, D. (1975). Why the law of effect will not go away. *Journal of the Theory of Social Behavior.* V, 2:169–87. [Page number taken from reprint in : Dennett, D. *Brainstorms: Philosophical Essays on Mind and Psychology.* Cambridge: Bradford Books (1978).]

Dennett, D. (1983) Intentional systems in cognitive ethology: the "Panglossian paradigm" defended. *Behavioral and Brain Sciences* 6:343–90. [Reprinted as Chapter 7 of Dennett, D. (1987) The Intentional Stance, MIT Press / Bradford Books.]

Eldredge, N. (1985) *Time Frames: the rethinking of Darwinian evolution.* New York: Simon and Schuster.

Ghiselin, M. (1974). A radical solution to the species problem. *Systematic Zoology* 23: 536–544.

Gould, S. J. (1982a) Darwinism and the Expansion of evolutionary theory. *Science* 216:380–387.

Gould, S. J. (1982b) The meaning of punctuated equilibrium and its role in validating a hierarchical approach to macroevolution. In: Milkman, R. (Ed.) *Perspectives on evolution.* Sunderland, MA: Sinauer Associates.

Ho, M. W., and Saunders, P. T. (1979) Beyond neo-Darwinism—an epigenetic approach to evolution. *Journal of Theoretical Biology* 78:573–591.

Hull, D. (1978) A matter of Individuality. *Philosophy of Science* 45:335–360.

Huxley, J. S. (1942) *Evolution: The modern synthesis.* London: Alan and Unwin.

James, W. (1890) *Principles of Psychology.* New York: Holt. [Page numbers taken from Great Books edition, University of Chicago Press, 1952]

Laszlo, E. (1981) Biperspectivism: An Evolutionary Systems Approach to the Mind-Body Problem. *Zygon,* 16, no. 2

Monod, J. (1971) *Chance and necessity: An essay on the natural philosophy of modern biology.* New York: Knopf.

Newell, A. and Simon, H.A. (1963) General problem solver: a program that simulates human thought. In: Feigenbaum, A. and Feldman, V. (Eds.) *Computers and Thought.* 279–293. New York: McGraw Hill.

Paley, William (1802). *Natural Theology: or, evidences of the existence and attributes of the deity collected from the appearances of nature.* Boston: Gould, Kendall and Lincoln, 1839.

Piaget, J. (1978) *Behavior and evolution.* New York: Pantheon books.

Reidl, R. Systems-analytical approach to macroevolutionary phenomena. *Quarterly Review of Biology* 52:351–370.

Rozin, P. and Schull, J. (1988) The adaptive-evolutionary point of view in experimental psychology. Atkinson, R. C., Hernstein, R. J. Lindzey, G., and Luce R. D. (Eds.). *Handbook of Experimental Psychology.* Wiley-Interscience.

Schull, J. (1990) Are species intelligent? *Behavioral and Brain Sciences,* 13:63–108.

Searle, J. (1980) Minds, brains and programs. *Behavioral and Brain Sciences* 3: 417–24.

Simon, H. A. (1962) The architecture of complexity. Proceedings of the American Philosophical Society, 106, 467–482. Reprinted in Simon, H. A. (1970), *The Sciences of the Artificial.* Cambridge: MIT Press.

Skinner, B. F. (1938) *The behavior of organisms An experimental analysis.* New York: Appleton-Century-Crofts, Inc.

Skinner, B. F. (1969) The Phylogeny and ontogeny of behavior. In: Skinner, B. F. *Contingencies of reinforcement: A theoretical analysis.* New York: Appleton-Century-Crofts.

Skinner, B. F. (1975) The shaping of phylogenic behavior. *Journal of the Experimental Analysis of Behavior* 24(1):117–120.

Thorndike, E. L. (1898). Animal intelligence: an experimental study of the associative processes in animals. *Psychological Monographs,* 2(Whole No. 8).

Valsiner, J. (1984) Conceptualizing intelligence: from an internal static attribution to the study of the process structure of organism-environment relationships. *International Journal of Psychology* 19:363–389.

Waddington, C. (1975) *The evolution of an evolutionist.* Ithaca, N. Y.: Cornell University Press.

Weiss, P. (1970) The living system: determinism stratified. A. Koestler, and J. R. Smythies (eds.) *Beyond reductionism: new perspectives in the life sciences.* New York: Macmillan 3–42.

CHAPTER 5

Modelling Biological and Social Change: Dynamical Replicative Network Theory

VILMOS CSÁNYI AND GYÖRGY KAMPIS

Editor's Introduction: The chapter below contains a concise summary of a theory propounded by Csányi in his *Evolutionary Systems and Society: A General Theory* (Duke University Press, 1989). The theory carries the enlargement of the concept of the unit of
evolution to its logical conclusion: that unit is now the replicative system *per se*. Its exemplifications range from the cell and certain subcellular units, to the biosphere itself. In
describing processes of evolution in reference to the replicative system, Csányi elaborates a
theory of evolution with a truly general scope.

The interest of this summary of the theory is two-fold: on the one hand, it presents a
comprehensive overview of its main tenets, and on the other, it shows how a theory developed primarily on the basis of processes observed in the realm of biology—self-replication,
autogenesis, convergence, etc.—can shed light on the processes of human and social
evolution.

Co-authored with Csányi's close collaborator György Kampis, the study below offers
clear-cut definitions of a number of basic evolutionary concepts, and raises a number of
fundamental research questions. These have an interest of their own, beyond the specifics
of this, or any other, particular theory.

As Bocchi has shown, in evolutionary biology there are heated debates
about the relevance of neo-Darwinian theory. Extensions and alternatives
are proposed, such as coevolutionary, hierarchical and non-adaptationist
models. However, there is little consensus about these newer developments. This is largely so because the arguments do not seem to refer to
any common ground. Based on this recognition, interest is now slowly
turning towards theories which provide a methodological construct for
their evaluation.

During the past few years, a general evolutionary theory has been elaborated by the authors and published in several papers and a monograph. This theory endeavours to bring the various bio-organismic levels of evolution into a unified conceptual framework, and include in it also the human cultural and technical levels, omitted in classical theories of evolution.

The theory of *replicative systems* predicts a number of previously unknown features of evolutionary systems, such as the direction of evolution, the mechanism of the emergence of organizational levels and compartments, etc. We believe that it may serve as the basis for a new paradigm for the social sciences as well.

THE REPLICATIVE MODEL
[DEFINITIONS AND DISCUSSIONS]

System

A system is a finite physical space separated from the background by topological or organizational boundaries, in which building blocks of physical *components* are present. The components are assembled and disassembled continuously by component-producing processes: there is an energy flux flowing through the system, capable of exciting some of the building blocks. Such a system is called *component-system*. The number and types of building blocks in the system, the energy flux, the material sources, etc., constitute the *parameters* of the system.

Function

The ability of components to influence the *probability of the genesis or survival* of other components of the system, due to their relationships with the component-producing or component-decaying processes, is called function.

Zero-System

A system of components which has not yet developed functions is called a *O-system*.

Information

Information is a specific description of components on the basis of the arrangement of the building blocks. Two principal kinds of information are distinguished:

Parametric Information

Components of a O-system carry structural information which is merely a manifestation of the system's parameters. This is called parametric information.

Functional Information

Components endowed with function carry information to the system, as their function is one of the determinants of the component-producing process. This information is related in some way to the arrangement of building blocks in the respective components, i.e. it is also structural information. The components' structural information, bearing function is called functional information.

Replicative Function, Replicative Information

The most important among the numerous possible functions is the replicative function. This term denotes an effect owing to which the *probability of the genesis* of the same component (or a set of components) carrying the replicative function *increases* in the system. The structural information of components which carry replicative function is replicative information.

Organization

The interrelated network of components and component-producing processes, i.e., the network of functions, constitutes the organization of the system.

Replication

Replication is an imperfect copying of the components, directed by information which is located either in the copied component itself or is distributed in the system.

Both the system and its components are produced in the replicative process.

In a *copying process* a constructor produces a copy (replica) of a component or of a system. The constructor provides the necessary information for the copying process. While the copying itself depends on the functional operation based on this information, it is independent of the particular mechanism of information storage and retrieval.

In a replicative organization components are endowed with functions, expressed by functional information. In a component-producing process, regenerating the system, functional information is also regenerated. The process is guided by the same information. Thus it is literally a self-copying process.

Two forms of replication are distinguished:

Temporal Replication

Temporal replication is the continuous renewal of a system or subsystem in time, by permanent, functional renewal of the components, while conserving the unity and identity of the system.

Spatial Replication

Spatial replication is identical to "reproduction": a system produces its own replica, which becomes separated in space. From one unit, two units are formed.

The same structural information is replicated in both temporal and spatial replication. In the temporal replication the *structure* of the system remains unchanged by the replication of components, while in spatial replication the structure of the system changes (it is duplicated). In both cases the system's *organization* remains unchanged.

Fidelity of Replication

Replicative processes are usually not error-free. The nature of replication can thus be defined by its degree of fidelity. If it is precise for all parameters, it is called *identical replication*. In case of non-identical replication, either the structure of the components or the component composition of the system changes relative to the preceding state.

Autogenetic System-Precursor (AGSP)

AGSP is a minimal set of components which is able to replicate and which fulfills the following criteria:

a. It contains at least one cyclic component-producing process.

b. At least one of the components participating in the cycle can be excited by the energy flux flowing through the system.

Operation of the Replicative Model, Autogenesis

Based on various considerations and on data concerning observed biological and social systems, we can say that the functional information content of a O-system containing an AGSP increases as its parametric information content decreases. This process is called *autogenesis,* an evolutionary dynamic predicted by replicative theory.

An increasing part of functional information becomes replicative information. This occurs by an extension of AGSP: that is, additional replicative cycles appear which are interconnected with AGSP. These for mations are *supercycles.* A replicative coordination of the supercycles develops, characterized by increasing fidelity of replication. Functional differentiation and cooperation appears, and it ultimately results in the formation of communities of simultaneously replicating components (i.e., sub-systems) called *compartments.* The components of these are separated on the basis of their co-replication. The emergence of compartments involves the organizational, i.e., functional, closure of the component-producing processes of components with replicative function. This is the *compartmentalization and convergence of replicative information.*

Compartments, already replicating with relatively high fidelity, can develop functional relationships to each other. Then a new component-system comes into being *on a new level,* of which the components are the compartments of the former organizational level. The fidelity of replication within the former compartments can be high, but on the next level, in the new component-system, it can still be low. On the new level a new autogenetic process can commence. As a result compartments of compartments come into being in the course of replicative coordination. Eventually the whole system begins to replicate as a unity, with more and more perfect fidelity on all its levels.

In the autogenetic process the organization of the system and of its parts changes due to the functions of the existing components. Thus autogenesis is possible only when the stage of identical replication has not been achieved. In the identical replication stage the system is functionally closed and its replication continues as long as the environment does not change. There are no further organizational changes initiated by organizational causes, because no new functions can originate. The system becomes an autonomous self-maintaining unity, a network of component-producing processes which, through the functional interaction of components, produces exactly the same network which has originally produced them. Its organization is closed and cyclic; its inputs and outputs are subordinated to its replication. Nevertheless, its existence depends on the invariance of the environment.

The notion of function expresses that autogenetic systems are not simply dynamical processes. Autogenesis is the evolution of active self-construction.

THE REPLICATIVE PARADIGM

Replicative Networks in the Biological World

The *living cell* is a molecular network which creates itself by producing its components in a replicative process: its only function is to produce itself. Information, necessary for its functioning is embodied in its molecular components. It is replicative information. The same applies to organisms and ecosystems.

The formation of higher and higher organizational levels is a consequence of the non-identical replication of the respective compartments which, due to their high fidelity of replication, are stable enough to serve as components on the next level; at the same time the inaccuracy of this replicative process provides room for further evolution. The driving forces of evolution are those functions which do not belong to a closed organization. In the long run the direction of evolutionary change is towards identically replicating organization on all levels.

These insights can be contrasted to the traditional view on biological evolution. There are two main tenets of the classical Darwinian evolutionary theory: the spontaneous, inherent variability among individual organisms and the mechanism of natural selection. Natural selection working on individual variability leads to adaptation. Many authors debated the adaptation concept recently, because several neutral or mildly harmful traits are found in various species which were not removed by natural selection. If we accept that several traits of an organism emerged independently of the adaptation process, then we have to suggest an explanatory mechanism for their emergence; the classical theory is inadequate for that.

The replicative theory provides such explanatory mechanisms. First of all, it does not restrict the evolutionary process to the level of individual organisms, but treats the whole biosphere as a systemic unity with several organizational levels in which each level is a theatre of evolutionary changes. Moreover, it operates with the concept of replicative selection besides competitive selection. Replicative selection is a highly creative mechanism which eliminates only those variants which are unable to replicate. The concept of replicative selection circumvents the tautology of fitness and adaptation.

Replication depends on the replicative information of the components and is independent from traits which do not contribute to or inhibit the replicative process of the given component. The concept of replicative selection eliminates the sharp dichotomy of adaptive and nonadaptive traits.

Replicative Entities in Social Systems

Replication of Ideas

Culture is usually defined as a set of learned, functionally related behavioural patterns and material products produced by them. These behaviour patterns are generated by a population of acquired neural codes in the brain.

We define *ideas* as physical representations of acquired neural codes which generate the smallest meaningfully definable actions or thoughts that have communicated, imitated, created as artifacts or performed as social behaviour. A complex social act, a thought and an artifact involve an *idea population*. Ideas, as physical entities contain structural information. Through their interactions they may have functions by which they can influence the probability of the genesis of other ideas. Ideas are created in a replicative process—imitation, transfer and reconstruction through language, etc., and the fidelity of their replication can be anything between 0 and 1.

There are three main categories of ideas: social, material, and abstract.

Social Ideas

Man lives in complex social interrelationships. Social behavior is molded by imitation, learning and discipline. In the brain of each individual a characteristic idea population is formed, which regulates his or her social actions. Ideas determine not only personal relations, social institutions and one's relations to these institutions, but also various value orientations. In case of a closed group the social structure of the group is formed by the idea populations in the brains of its members. Determination of social structure is populational and stochastic in character.

The individual as a carrier of a segment of the idea population that determines social structure is a creator of this very structure, one which is normally his social environment.

Material Ideas

This category contains ideas known through the creation of artifacts. Even simple objects of everyday life represent ideas: ideas expressed in the function, in the value, or in the way of production of the artifacts. During the creation of an artifact the object itself becomes a carrier of structural information. The idea population in the maker's brain obtains a characteristic representation in the artifact; it appears as structural information in the object. This information-transfer makes it possible to return the information located in the object to the brains of humans. Hence, human brains and the world of artifacts are two continuously communicating compartments of the idea domain. These interrelations can be readily observed during various social rites, in which the presence of artifacts contributes to the formation of proper social behaviour. The objects used in the rites help the participants remember, perform and transmit the special sequences of behaviour. Fantasies and artifacts lead to an exchange of information and the development of interwoven functions.

There are many signs of the separation of the autogenesis of the information represented by artifacts (technical evolution) within cultural evolution as a whole.

Abstract Ideas

Those ideas belong to this category that are represented neither in artifacts nor in social relations. For example, the ideas of a story are abstract ideas, but the tale can be written or printed and then these copies operate as artifacts; the tale may contain social references and therefore it may have an influence on social interactions. Abstract ideas, replicated by spoken language, may spread and propagate quickly. For example, the abstract idea of a good joke can reach millions of people in a city in a couple of days. Mass communication has made idea replication even faster.

Networks of Ideas

Ideas of the above *idea compartments* are *components* bearing *structural* and *functional information* and their system is a *component system*.

In human behaviour ideas are replicated in two ways. First, they are replicated in a given human population. They spread among people with remarkable fidelity. Second, they are transferred from the one to the next generation. Educational systems are organized primarily to provide anappropriate milieu for idea replication of high fidelity. Different ideas may compete and cooperate with each other because their physical space, i.e. the space of human brains, is limited. An entire idea system can be regarded as a collection of *idea components* and component-producing processes. This statement is valid both for the individual and for large groups.

In the case of the individual a regulated network of ideas is constructed during development and maturation. This network of ideas consists mostly of functionally connected ideas that supplement each other and represent a kind of "ecosystem" which behaves as a unity in many ways. The network replicates itself in time; its organization and structure become more and more mature and rigid during aging. Part of this system may be replicated in social interactions. It is important to note that people spend most of their time (besides sleeping) in *talking*, that is in exchanging and copying ideas.

In smaller groups the limited idea-space of the group becomes quickly saturated and a more or less closed network of ideas develops. This is a newly emerging entity. It is a system of components (ideas) and component-producing processes. It is replicating in time and occasionally also in space. Its existence becomes independent of individuals because the idea populations are shared in different individuals and generations. The creator of an idea may have been dead for hundreds of years but the ideas he invented may still be replicated within the group. By this separation of ideas from their creators a new level of evolution comes into being: the level of *cultural evolution.*

Society as a Complex Replicative Network

The Act of Creation: Creative Spaces

Society can be regarded as a complex component-system. Its components are human beings, ideas, artifacts and living species exploited by man. The creation of a component involves the action and contribution of other components. Creation of a complex object or idea requires the contribution of a fairly large set of components. To create a particular-object, some raw material is needed; to get the raw material, other objects are needed, and so on. We use a concept which reflects the complexity of the creative process: *creative space.*

The creative space is a model, an abstract space in which a representation is given of every component involved in the production of other components. So we speak of technical space, i.e., the creative space of artifacts: of cultural space, which creates various political and social ideas, and biological space, which creates living beings.

It can be shown that creative spaces have *replicative organization:* the information represented in the created components feeds back to the processes which it created.

For example, cars are created in factories where the relevant information is in blueprints. The information transfer from blueprints to artifacts is a seemingly one-directional process. But if we examine the entire creative space of artifacts, it appears that the already manufactured cars *influence* the car-producing process as well as the blueprints. The most common process of design is *copying:* the designer copies, and sometimes recombines parts of the cars made in previous production cycles. Even if he invents something, the basis of his invention could have originated from the creation of a different artifact. Obvious examples are the classic cars which resemble horsedrawn carriages in many ways. The fidelity of replication is not always perfect—it does not exclude changes, inventions in the processes of creation.

The same process can be observed in the creation of various ideas. A formed, written or composed idea may be thought of as the product of the creative mind. But, if we consider the entire creative cultural space, many other ideas can be identified that provide the necessary information for the creative mind. Ideas are copied, recombined and only to a lesser extent newly invented; the latter is defined by the fidelity coefficient of replication.

The replicative processes of the other components of society, namely human beings and other biological organisms, are evident and need no further discussion here.

Boundary Conditions of Creation

The act of creation in the creative spaces is influenced by many factors other than replicative information. For example, living beings are created within the boundary conditions of the processes of life, but with the advancement of biotechnology these boundary conditions are now rapidly changing. The creation of ideas is affected by the biological nature of man, so it would be important to account for its limitations. Also the creation of artifacts is influenced by chemical and physical boundary conditions.

System Generating Processes: the Emergence of Dynamical Replicative Networks

Up to this point we have dealt individually with the creative processes of the different component categories; now we emphasize their inherent interrelationships.

A complex component-producing system emerged on the surface of Earth that persists and behaves as a *systemic unity:* we shall call it the *Global System.* It produces living beings, artifacts and ideas. AGSPs emerge in the Global System spontaneously and continuously.

As a consequence, there are many simultaneously developing simple and complex replicative compartments organized around the initial AGSPs on Earth. These compartments influence each other partly because they are in competition for resources (energy, material, human minds, etc.) and partly because the operation of replicative information leads to the emergence of several organizational levels (for example in the social sphere: bands, tribes, cities, states, pacts, etc.). The result of this autogenetic chain reaction is a *dynamical replicative network* of the various components and component-producing processes.

Replication is regulated by feed-back loops and other cybernetic devices. The various compartments may *fuse, divide, include each other,* and they may functionally *differentiate.* It follows from the nature of replicative information that it extends in the system. Increasingly united compartments are formed and, as its most important feature, the process converges towards a balanced, high fidelity replication of the entire Global System.

The emerged replicative compartments (states, nations, cultures, industrial complexes) originating in the interactions of the components can co-exist only if they develop *functional* connections to each other. Through hierarchical replicative organization all organizational levels and compartments in the network are functionally interdependent. These functional connections are subordinate to a common, regulated global replication of increasing fidelity. The mutual regulation of the compartments by the replicative process occurs through competition, sometimes through cooperation, and occasionally through a drastic reorganization or "catastrophes." Inevitably, an organizational hierarchy forms at an increasing rate, and for each compartment active participation in the common replication process becomes the condition of persistence.

RESEARCH PROBLEMS

Description of Real Replicative Networks

There are diverse methods, practical tools, and concepts at the disposal of social scientists to describe the cultural, economical or ecological states of their subjects as groups, states or cultures. How can these concepts and methods be used in the framework of the replicative paradigm, how can one identify actual replicative cycles, or networks? More specifically, how can one determine the replicative fidelity of the components and of organization? Can existing data be used, or does the inner logic of replicative models require new types of data as well? How can a complex highly developed society be divided into compartments?

The most important question is whether the evaluation of the data on the basis of the replicative paradigm would result in a better understanding of social and societal processes. Expressed in the most general terms, these are questions of parameter identification and of methodology.

Application of the Replicative Model to Historical Cases

The applicability of the replicative framework should be checked against cases where complete and well-documented data are available. To this end, first one should develop a replicative model of a carefully chosen historical case. Testing the model would show whether corrections are necessary in the theory, and in the practical approach. In case studies one should show how a concrete organization works, and how the conditions of replication are to be identified. Taking this as a starting point we can infer the possible outcomes of the given situation in the replicative model, and can match these against actual historical data.

A first task is the interpretation of historically known causes of crucial changes or transitions in the replicative framework. Then the raw data is to be processed on the basis of this framework, and possible new or pre-existing instabilities are to be identified. How can these be generalized? What are the crucial individual functional relations and why; or is the entire organization unstable in a transition? For example, some functional relations may be too rigid and specialized, or the organization may depend too greatly on such functions.

The theory of autogenesis predicts that, when close to the organizational end state, the external stability of the system greatly diminishes. What is the mechanism of this, how can internal and external effects be separated, and what are the limits of internal control in concrete cases?

Application of the Replicative Model to Modern Social Crisis and Natural Catastrophes

Drastic social changes can be studied as well, where the organization may not only suddenly change, but may be seriously damaged or entirely destroyed.

Catastrophes in Replicative Networks

A catastrophe is a sudden, uncontrolled appearance of an effect, leading to major changes in the system (although this definition contains some subjective elements). Informational and dynamic catastrophes may be studied. A *dynamic catastrophe* is understood as physical damage of the networks caused by some outside force or by inner parameter-fluctuations in the sense used by R. Thom. A replicative system is a dynamical system, therefore it may show dynamical instabilities the same way as other systems. It should be noted, however, that even this form of catastrophe cannot be examined independently of the special properties of replicative systems.

Informational catastrophes are the results of the actions of active information carriers which interfere with the replicative cycles of the social system, upsetting its organization. They appear either due to an instability of organization, or as a result of uncontrolled outside effects which make their way directly through the informational processes.

In both cases the catastrophe leads to functional changes which interrupt or destroy parts of the organization. This means a change in the *identity* of the system, because identity can be meaningfully defined only through organization. Those functional relations of the disturbed organization which no longer participate in replicative cycles cause (or may cause) changes affecting the entire organization through chains of functional relationships.

Stability, Resilience and Learning in Replicative Networks

It is of great importance to study controlled changes in replicative systems. *Stability* is the ability of organization to keep important characteristic quantities in a given range through cybernetic processes within the system. *Resilience,* on the other hand, is the ability of organization to recover after changes.

A question complementary to the mechanisms of catastrophes is to be examined: the range of identity independently of what happens *beyond* it. More precisely, what are the conditions of stability and resilience, both in organizational and dynamical terms? What are the determinants of stability—is it related to the main material transfer in the social system or do more subtle informational relations play a role as well? As to resilience—what characterizes replicative social organizations having this property?

One kind of controlled change means *learning* in a network. Learning is a change of organization due to the external functional relationships of a replicative social compartment. How can we change replicative networks in a given direction and towards a given organization? To what extent is learning possible and what is its mechanism? This point is obviously important also from a practical point of view.

Effects of Hierarchical Organization in the Replicative Process

There is an organizational hierarchy in every replicative system, this results in the nested, "chinese box" nature of replicative compartments (such as cell-organism-ecosystem). Unfortunately it is not known how the operation of the entire system is influenced by regular and irregular changes of its various hierarchical levels. Which organizational levels must be manipulated and how, if changes are to occur in certain directions at given levels?

Another important question concerns the detailed conditions of co-replication and cooperation on different levels (e.g. in terms of the ratios of rates of organizational change on the levels). A further question is how the levels control each other, and how this is related to stability, resilience and catastrophes. For example, there can be an increase of stability due to the interaction of multiple level processes, or there can be 'cross-catastrophes' in which changes on one organizational level lead to catastrophes on another.

Natural catastrophes and their effects on the hierarchical replicative organization can also be studied.

Then we come to the question, how new organizational elements can appear in the multi-level system due to external control in the prevention of catastrophes. Also to be examined is the role of ideas (education) and of their reproduction.

Network Precursors in Societal Changes

According to our theory, larger system-reorganizing changes can take place as a result of the action of AGSPs. These are directed changes because as the precursors are activated, the reorganization takes the rules of the current organization and the AGSP as a starting point.

In case of existing societies it is also possible to study how such small reorganizing centers appear or have appeared in the past, and whether such a mechanism is feasible. If it is, what are its characteristics? How and where within the organization are precursors formed? Can they be recognized reliably before they start to act? What are the degrees of freedom in reorganizing a replicative network?

A Replicative Model of the Global System

On the basis of both theoretical and practical projects, it is possible to simulate the Global System with appropriate simplifications. The main functional connections and replicative pathways need to be identified, taking into account the interdependence of different organizational levels. The basic question is how the ultimate replicative state is reached in human society.

Will this state be influenced by energy and material resources? How resilient is a global social system in relation to sudden resource shifts? Can a perfect recycling technology be achieved? The possible forms of the global replicative equilibrium need also to be studied. For example, we can envisage two very different modalities:

a. at the point of equilibrium every society develops to nearly the same economic level, or

b. there are large differences in degrees of development at equilibrium point.

These alternatives entail major economical, political, and moral consequences.

BASIC BIBLIOGRAPHY

Csányi, V. 1982: General Theory of Evolution, *Publ. House of the Hung. Acad. Sci.*, Budapest.

Csányi, V. 1985: Autogenesis: the Evolution of Self-organizing Systems, in: *Dynamics of Macrosystems* (J.-P. Aubin, D. Saari and K. Sigmund eds.), Springer, pp. 253–267.

Csányi, V. 1989: *Evolutionary Systems and Society: A General Theory*. Duke University Press, Durham.

Csányi, V. and Kampis G. 1985: Autogenesis: the Evolution of Replicative Systems. *J. Theor. Biol. 114*, 303–323.

Kampis, G. 1987: Some Problems of System Descriptions I.: Function, *Int. J. Gen. Systems 13*, 143–156.

Kampis, G. 1987: Some Problems of System Descriptions II.: Information, Int. *J. Gen. Systems 13*, 157–171.

Kampis, G. and Csányi, V. 1987: A Computer Model of Autogenesis, *Kybernetes, 16*, 163–181.

Kampis, G. 1990. Self-Modifying Systems in *Biology and Cognitive Science: A New Framework for Dynamics, Information and Complexity*, Pergamon, Oxford, to appear.

CHAPTER 6

Social Evolution: A Nonequilibrium Systems Model

ROBERT ARTIGIANI

Editor's Introduction: A historian by profession, Artigiani undertakes here to address the question posed by Prigogine in the opening lines of his Foreword: to wit, To what extent do human activities stand outside the natural world? Prigogine asserts that at present our conception of nature is undergoing a radical change; a new paradigm is taking shape. Artigiani shows what that paradigm is when applied to the field of historical analysis.

Beyond mere metaphors, this study advances the "strong hypothesis" that human societies are examples of dissipative structures, their evolution accurately describable using elements of that model. It concludes that the model can both assimilate historical data and suggest a new framework of interpretation that deepens our understanding of key events. These are big claims, and Artigiani admittedly cannot fully substantiate them in the space of a single chapter. He does, however, analyze the emergence of civilizations in historical development, and shows that concepts such as fluctuations, bifurcations, and the emergence of order in nonequilibrium systems have meaningful applications to processes in human societies. Civilizations appear as basic phase changes that transform both the relationships that determine societies and the behavior patterns of their members. The analogy with biological speciation is striking. If the latter is discontinuous and relatively sudden (cf. the current debates discussed by Bocchi), so is the process of social change. Hence an important new element is added to our understanding of evolution in the human sphere: the element of saltatory, nonlinear and discontinuous change. Societies, like biological systems, maintain themselves by processes of replication (as underscored by Csanyi), but they also evolve through phase changes that amount to bifurcations in dynamical systems (as defined by Abraham).

If the basic concepts of replicative systems apply to the persistence and spread of societal structures, and if the nonequilibrium system bifurcations of dissipative structure theory apply to processes of historical change and development, we may have the basic elements of the conceptual foundation required to overcome the universally regretted but seldom effectively combatted "two-culture gap." We could begin to see how we might overcome the still influential positivist conception of history without the questionable benefits of *ad hoc* postulates and metaphysical principles.

Uniting quantum theory and irreversible thermodynamics, contemporary science attempts to track the creative power of nature [Davies, 1988]. The result is a paradigm of nature evolving, a nature of Becoming rather than Being [Prigogine, 1980]. In a nature where the process of Becoming replaces the illusion of Being [Rae, 1986], randomness, structure, and time account for qualitative change. Randomness provides a "cloud of evolutionary possibilities" [Allen and McGlade, 1987], some of which may trigger nonlinear processes driving systems through symmetry-breaking transitions to ever more complex levels of spontaneously self-organized structure [Prigogine, 1982]. Since increasingly complex structures are sustained by greater rates of external entropy production, the effect of the Second Law makes the evolution of complexity irreversible. Nature therefore has a direction in time, and, over time, qualitatively new levels of reality may be created [Chaisson, 1988].

The power of this scientific model invites speculation about its limits. At present, they are unclear, although the new paradigm has been used to model the origins of life [Corliss, 1988]. But the emergence of qualitatively distinct human characteristics has eluded scientific description. Using the emergence of "consciousness" as an example of qualitative change through emergence, the following is a hypothetical exercise that seeks to demonstrate the kind of theoretical model needed. Taking the biological foundations of consciousness for granted, I hope to model the nonequilibrium conditions in which its distinctively human form was achieved.

The model is designed to spark debate, and I do not pretend that all the steps described really took place. But I believe the arguments in this paper are typical of what will be necessary for the convergence of the human and natural sciences [Laszlo, 1987]. Of course, social scientists prefer an empirically verified description of actual events, and they view heuristic exercises suspiciously. But to know what events to look for, we must know what a nonequilibrium model of social evolution looks like. This paper attempts to aid in visualizing a difficult and complicated process.

OVERVIEW

"Return with us now to the glorius days of yesteryear" when social theorists could postulate a state of nature. Imagine the simplest conceivable human groups. They would already have the capacity for language, technology, and culture, of course, since these are anticipated in other primates. But for the sake of visualizing the process of social evolution, suppose these idealized groups, in relative terms and for modelling purposes, near-to-equilibrium. Consider them loosely ordered, small, and widely dispersed.

At equilibrium, the state of maximum entropy, there is no structure. All is lost in a meaningless, disorganized uniformity, and every motion is counter-acted by one in the opposite direction. Actions are random, and there is no "information." Neither place, time, nor direction can be identified. Every state being equally probable, momentarily appearing configurations do not reduce an observer's uncertainty about their future. No existing state makes its successors more or less likely. The timeless symmetry of the equilibrium world is broken when nonlinear processes involving positive feedback loops [Ulanowicz and Hannon, 1987] occur. Then energy flows "load the dice," creating propensities [Popper, 1982]. Randomly generated nucleations in a bounded context can now be amplified and stabilized [Glansdorf and Prigogine, 1971]. Far-from-equilibrium, structures can self-organize when their internal entropy is less than the entropy deposited into the environment [Nicolis and Prigogine, 1977]. Because the self-organized system is distinguishable from its environment, it communicates information. Its stability and structure reduce uncertainty about the nature of the world, and observations of an existing state permit predictions about its future.

It would be difficult to distinguish one hypothetical near-to-equilibrium group from another, or any from their environment. Because all arrangements of the people in these theoretical primal groups were almost equally probable, knowing where members are at one moment indicates very little about where they will be at the next. For millenia these groups wandered across a natural landscape on which they made little impression. They built no permanent structures and carried few belongings. Their possessions were small, light, and perishable. Few jobs needed to be done, and they were usually done by individuals indiscriminately. Although biological bonds no doubt existed, since the groups contained gregarious creatures, hierarchies and societal control mechanisms were limited. These wandering groups, in fact, would have had only a vague sense of their collective identity, and their structuring information was too minimal to distinguish between tales, traditions, and legends. Nor would their languages have reached the level of abstraction permitting symbols to be logically related [Cassirer, 1946].

The earliest human groups oscillated over a narrow range of stable states. When destabilized by fluctuations in their environments, they usually collapsed toward equilibrium. Any small band wandering into a new environment, for instance, multiplied briefly because of abundant resources. When population saturated the resources available, however, a bifurcation took place. The group had to choose between disintegration and organizing its members cooperatively in an intensive effort to sustain increased population by accessing new energy resources. Lacking the capacity to organize, the group either died off or fragmented, dividing into small parties searching anew for virgin lands. Under these conditions, a balance between human population and geographical territory was achieved. A strata of humans, gathered in small, loosely organized and widely dispersed groups spread thinly in space.

With the appearance of civilizations in a few river valleys about five to six thousand years ago a qualitative change in the form of social organization occurred and the symmetry of early human existence was broken [McNeill, 1963]. Organized structures of much greater complexity suddenly emerged that radically distinguished themselves from the smaller, simpler, nomadic groups that had previously characterized human experience [Meggers, 1954]. It is obvious that the valleys themselves played a crucial part in this process. They were alien territories that, apparently, people had originally avoided. But population dispersion had gradually used up all preferred sites, and stragglers had no option but to explore these tangled, damp, malarial locales.

A surprise awaited the new arrivals, however, for in the valleys they found environmental resources able to amplify alternative ordering possibilities. Water and silt made possible more enduring settlements on a larger scale. Floods regularly irrigated fields and made crop yields larger; silting renewed soil fertility and made higher yields predictable. In an energy rich environment, wanderers could become sedentary, and, with some coordinated effort, they were soon more prosperous and numerous than surrounding nomads. The river valleys, in other words, supplied energy that drove social systems away from equilibrium, enriching the information content of human groups and making proto-civilized societies increasingly distinct from nature. Obviously, variations in local conditions stimulated diversity between the social structures developed in each settlement, making sedentary societies increasingly distinct from one another as well [Frankfort, 1948].

It only remained for proto-civilized societies to preserve their highly improbable states. Systemic factors were at hand to establish new boundary conditions and institutionalize social structures with division of labor, hierarchy, and increased interdependence. Increased wealth attracted attention from threatening nomads, for instance, whose attacks provided a gradient preserving nonequilibrium conditions [Reilly, 1989]. Building protective fortifications, in turn, clarified boundaries and stimulated the growth of cities, tying people to permanent settlements [Carniero, 1970]. Similarly, social cooperation intensified the exploitation of resources by developing internal constraints. Storage facilities and distribution systems reduced internal entropy production and made possible successful transitions towards higher levels of complexity. The expanded fruits of forced labor were gathered together and distributed inequitably, increasing the energy extracted from internal flows by allowing the privileged few to coerce even greater sacrifices from the exploited many. As a result, ever more complex, clearly defined social structures organized themselves.

ENTROPY AND SOCIAL EVOLUTION

The earliest human groups were looser in their internal cohesion, more uniform in their behaviors, and less complex in their structures than the proto-civilized societies developing in the river valleys. Nevertheless, every evolutionary departure depends upon the reservoir of energy accumulated through the historical development of separate component parts, which, when united, produce a qualitatively new emergent form [Csanyi, 1987]. Unless the parts had ratcheted themselves at least some distance from equilibrium, no platform for continued evolution would be available. Biological evolution obviously must have laid the foundations on which big-brained, bipedal humans could create their unique forms of society and consciousness. The problem is to find the sources of proto-civilized organization and the elements available catalyzing the evolutionary leap. An explanation based on thermodynamics must focus on the role of increased entropy. Increased external entropy production is one way to measure increased complexity, although the fact that the entropy deposited into the environment increases tells us little about the organization of a new structure. The loss in specificity, however, is more than compensated for, humanistically, by the indeterminism of the thermodynamic model.

Entropy increases associated with the evolution of emergent structure take place on two planes, one contingent and one principled. A contingent entropy burst, a sudden increase in systemic confusion that precedes evolutionary growth, is always associated with the transition to a higher level of organization. Contingent entropy bursts are results of unpredictable energy fluctuations in previously unperturbed stable structures that have grown rigid [Holling, 1986]. Stabilizing processes having become brittle, adaptability declines. That is, stable structures gradually lose their ability to adjust to fluctuations in environmental resources [Ulanowicz, 1986]. When environmental dynamics confront them with new circumstances, they may disintegrate. In effect, contingent entropy is a function of individual structural experience.

But evolution must conform to the principle of the Second Law. When a higher level of order emerges from local chaotic conditions, it is achieved by streamlining internal processes. Streamlining reduces the entropy within a structure. But work must be done to preserve the organization of internal processes. In more complex structures, there is more work to be done, and it requires the use of greater amounts of energy. Once used, energy must be dissipated into the external environment, and since more work is being done by a structure that has moved farther from equilibrium, energy dissipation—or external entropy production— is increased. Purchasing increased local complexity at the expense of increased global entropy means that evolution is an irreversible process in which the entropy in the universe increases.

Both contingent entropy bursts and structural entropy flows followed upon primitive societies entering the river valleys. Their unfamiliar environments, of course, destabilized precivilized structures, driving immigrant groups through highly entropic conditions to a bifurcation point where they chose between new behaviors and total collapse. Amidst confusion, a variety of behavioral alterations were stumbled upon, but only after much wasteful and contradictory experimentation. Eventually some of the altered behaviors interacted in a mutually enhancing positive feedback loop that drove societies toward more ordered states. But stabilization in a manner that took advantage of river valley resources meant more work was done by individuals and social systems grew larger and more complex. Work consumes energy, and the more work a social structure performs the more entropy it dissipates into the environment.

To preserve the information describing their more complex states, river valley societies developed new mythologies [Coulborn, 1969]. Propounded by charismatic leaders, new mythologies captured information about more ordered states. Passing beyond cosmogonic myths that registered group awareness of the external world, societies became so distinct from their environments and from one another that they became aware of themselves and needed to explain the origins of their own existence [Eliade, 1975]. To replicate their unique structures in future generations, the new myths symbolized the rigorous organization which was one strategy for responding to far-from-equilibrium conditions. By elevating a single male god—usually associated with a conquering city—above the others [Frankfort, 1973], the new myths focused attention on preserving stability. Allegiance to the society was intensified because the divinities were identified with it. When the gods became symbols of the state, societies had begun mapping themselves along with their natural environment. With the appearance of "state-gods" the language describing social order had emerged. Elevating myth and religion above legend and tale, social algorithms had gained "logical depth" [Bennett and Landauer, 1985]. That is, despite the tremendous work and terrible suffering accompanying the entropy burst in which river valley societies self-organized, inheritors of the new descriptive algorithms could preserve their societies efficiently and in relative safety.

WAR AND INFORMATION

If noise can provide ordering information in the rest of nature [Atlan, 1985], war could help produce information structuring societies. War, like trade, is simply a means for exchanging energy, matter, and information between human groups [Dixon, 1976]. But war allows us to see the effects of exchanges more clearly. Any release of matter, energy, and information by one society into another amounts to an observation of it. By observation, entities are identified. War, however, is an intrusive form of observation; its effects shape a society according to war's own structural predelictions. By linking civilization to war it is possible to understand how civilized societies became associated with many of the injustices attendent on the last 6,000 years of human development.

War helped define civilized societies in ways whose morality we now question. But, by the same token, war helped define civilization itself; it stimulated human systems to evolve from loosely ordered, near-to-equilibrium groups into highly organized, far-from-equilibrium societies. Violent attack was an observation that redefined an aggregation of more-or-less independent "yous" and "mes" into an integrated "us." It was by attack, most obviously, that what happened to one happened to all: Violence was a collective experience. An attack drove a surviving group farther away from equilibrium, selecting from the random distribution of people in the group one improbable arrangement, and violence made maintenance of the selected ordering dependent upon the interactions of its human parts. Through war nature correlated human behaviors and evolved a new level of societal reality.

Ancient warfare, according to Arther Ferrill [1985], originates with "formation." Tying war to formation, Ferrill fears, will seem a "grotesque simplification" to conventionally trained historians. But from an evolutionary point of view, his is a sophisticated thesis that irreversible thermodynamics can model. Organized through random events, "formation" reduced the internal entropy of human groups, moved them farther from equilibrium, provided access to greater energy, made them more complex, and laid foundations for the symmetry-breaking transition to civilization. Formation organized group behavior. Organization meant greater efficiency in accessing and using energy [Durham, 1976], providing the operational base necessary to evolve in the altered conditions found in the river valleys. Finally, at a proto-civilized level, war increased complexity by distinguishing between different types of violence and by discriminating between social roles. Violence became organized and purposeful, while individuals whose sole job was to fight became parts of social systems. "Formation" also created a command level in the social hierarchy that increased the flow of information, the amount of work done, the entropy produced, and the interdependency of social life.

Originally, there were only two military formations: the column and the line. Both are simple, and other animals hunt using something like them. The column is the formation of soldiers in a file, front to back, used to transport combattants to battle. The line is the arrangement of soldiers side by side actually used to fight. Both could be stumbled upon easily, yet, in the most basic sense, column and line represent increased information. Literally reducing uncertainty about the world, an observation of the men in column and line at one point in time allows for meaningful predictions about their futures. Those men are not randomly distributed. They are in highly improbable states with great mutual dependency, for predictable attacks made the survival of members of a formation more dependent upon one another. Interdependency means column and line are social phenomena: they express information about the collective world. The relationship between individuals, not the people themselves, defines column and line.

By controlling its members through military formation a society reduces internal entropy, gaining access to and processing resources less wastefully than its neighbors. This can be an extremely subtle distinction, which is why viewing society from the perspective of warfare has an advantage. It is obvious, for instance, that a column is more efficient at moving men than a gang. The random sequencing of a gang means that fewer of its members arrive at a destination than those in a column, and the gang makes it impossible to know in advance which members will actually show up. Individuals in a gang could wander off, sleep, go fishing, or lose interest in the enterprise. The column meant all raiders actually reached their goal. Use of formation, therefore, gave raiders an important advantage in the exchange of energy, matter, and information with competing groups. To preserve the advantages of formation strategies, societies learned to act purposefully.

Ferrill was right to worry about associating such a multitude of consequences with a single factor like formation; the selected factor acquires so great a burden that it is hard to imagine primitive peoples appreciating its potential. But, based on randomness and nonlinearities, a thermodynamic model would not claim the first formation raiders aimed to transform the nature of society. Instead, the model suggests primitive military formation was an accidental discovery that, interacting with environmental factors, willy-nilly moved human groups farther from equilibrium. War increased access to environmental resources, which interacted with formation to produce the symmetry-breaking positive feedback loop leading to civilization [Bronowski, 1973]. Formation thereby biased change in the direction of evolving complexity. It is an example of how self-organizing processes cooperate with natural selection in accelerating the evolution of complexity [Kauffman, 1984]. Thus, inventing formation would not necessarily involve prescience—or even conscious choice. Only a little bit of luck was required. In this case, luck should have been easy to come by; it may have been inherent in randomly generated initial conditions.

The nonequilibrium state in which column and line were discovered can be modelled readily. A gang of attackers would itself constitute a move away from the near-equilibrium of prehistoric life. A whole group would not attack, only some of its males would. The attack, of course, could have been motivated by impulses far removed from a thermodynamically justified search for expanded resources. Fits of pique generated by totally irrelevant factors could have triggered aggressive behavior. Nevertheless, the attacking process would accelerate the move away from equilibrium. Moving through enemy territory, probably along a woodland track in the dark, the attack party would have been made up of men with varying degrees of experience and native courage. But they would not remain randomly mixed. The veterans and the foolish would have moved ahead, while the rest, individually, would have struggled with their fears and uncertainties as they hurried to keep up. Distinctions like these would have been insignificant or mutually canceling in the near-to-equilibrium conditions of normal life. But in the agitated state accompanying an attack, individual distinctions became meaningful and, like droplets in a mist, a ragged column formed. Equally important, once formation had been stumbled upon a nonequilibrium condition between groups existed. Thus, the thermodynamic conditions essential to a

continuing exchange of energy, matter, and information within and between groups was created.

It is equally easy to postulate the physical conditions in which the line
was created. When the attacking column encountered the barrier of an
enemy camp, raiders drew up next to one another and the column was
transformed into the line. An attack mounted from this position—a direct sweeping motion—left vast avenues of escape. To prevent that,
early attackers used one military tactic, the "surround" [Turney-High,
1971]: They simply encircled the village to be victimized before assault-
ing it from all sides. This tactic was probably adapted from the ancient
hunting practice—already anticipated by lions and wolves—of spreading
members of the society out to drive game toward prepositioned killers.
The basic attack formation could have been based on hunting tactics, but
it was modified by the energy flow of the column and the obstacle imposed by the camp. The unpredictable use of an available behavior for
new purposes in a new situation illustrates the evolutionary strategy
Jacob [1982] calls "bricolage," or "tinkering." A structured form
emerges by "making do" with whatever components happen to be available, redefining inherited attributes by putting them to new uses. Thus,
the surround was differentiable from hunting, revealing an increase in
the complexity of group behavioral repertoires.

Column and line are highly improbable states, and the presence of
nonequilibrium conditions can account for them. At a primitive level of
development, the column could easily have appeared in a manner analogous to the convection cells developed in a Benard Instability. In this
case, a heat gradient rises upward through horizontal layers of a fluid in
a gravitational field. Because the heated layers are less dense, their molecules rise. At the top levels, however, they cool and fall downward.
When heat is first applied, all molecular motions are random and energy
passes between molecules by conduction. But if the proper boundary
conditions are reached—if, that is, the difference in temperature between the bottom and top layers achieves a critical threshold—there is a
"phase change." Random motion gives way to organized, hexagonally
patterned cells in which convection occurs, and energy is transferred
more efficiently.

Men are not molecules, but if, in a raid, excitement and fear affected
behavior over a distance great enough to manifest meaningful distinctions between levels of courage and experience, a chaotic gang would
have been transformed into an ordered form. Columns would then be
examples of self-organization that emerged through physical actions in
specific boundary conditions. More traditional approaches prefer credit-

ing human innovation with forethought and planning, with, in a word, consciousness. But Jaynes [1976] makes it clear that consciousness is an acquired human characteristic, not an inherent attribute. The attribute cannot be used as a cause of its own emergence. Of course, consciousness could have appeared earlier than either Jaynes's or this model suppose. The question here is can thermodynamics, with its explicit stochasticity, explain the appearance of conditions in which structures became complex enough for consciousness to emerge. Exactly when and how that happened is of less immediate significance.

ORGANIZATION

Formation made aggression efficient. It distinguished raiders from mobs, combat from riot. But preserving the newly won far-from-equilibrium structure obliged groups to organize themselves, formally constraining the connections between constituents by rules designed to stabilize an ordered state. The problem is to find the peculiarly human organizational patterns by which information generated at an instability is transferred to future generations.

Essentially, order refers to the establishment of improbable states. These result from nonequilibrium conditions, from a flow of energy, for the appearance of structure represents a set of meaningful relationships that would be damped at equilibrium. The accidental discovery of formation reported the frighteningly dangerous nonequilibrium conditions in which it first appeared, as the unique configuration of a crystal reports on the chemical conditions in which it precipitated. Nature records these symmetry-breaks historically when information is captured in matter. Matter writes time into nature by physically transporting information from the irretrievable past into the present. Early social structures may have been preserved in homologous ways. Amerindians moving toward a village selected for pillage held each other's hands; they preserved proven past behavior by organizing a column physically. Sooner or later, rules developed sustaining the ordered behavior morally, which required symbolic language. The process by which symbolic language emerged was stadial, depending, in its earliest instances, on the body as much as the brain, on gestural as much as verbal communication. When the information directing successful raiding grew too complex for gestural communication, a more abstract language capable of communicating ideas and values through verbal symbols was needed. But how was the essential information to be separated from the confusion of actual existence?

Success having broken the symmetry of previous experience, raiders may have reinforced the memory by reenacting their attack, reminding each other and themselves of their experiences. Reliving a successful behavior is the most elementary way in which a social group self-organizes, for iterations of the same behavior pattern experience. Play also provides opportunities for repeating and clarifying processes, making them more refined, efficient, and ordered [Callois, 1961]. Play acting, in turn, could easily become "dance," at which point its repetition constitutes a ritual [Harrison, 1924].

Ritual is "quintessential behavior" [Turner, 1969]. Isolating an experience from everyday events, ritual provides the opportunity to both reinforce social structure by repeating its constituent behavior and to educate youth. Ritual also supplies the mechanism for shifting from one social state to another [Turner, 1986]. Thus, ritual war dances preserved the structural order of a raid at the same time that they smoothed the transition from peace to conflict—imitating their ancestors, men could excite themselves to violence, form columns and lines, establish their mutual support, allay their fears, and dance their way into battle. Finally, ritualized repetitions also induce ecstatic states [Campbell, 1983], which may have produced the psychologically nonequilibrium conditions that helped cross the barrier to symbolic representation. Repeated in a trance-like state, experience could be simplified and captured in words, after which it can be integrated into the mythic lore, the algorithms guiding organized social behaviors.

Societal algorithms contain symbols exciting the repetition of recorded behaviors in appropriate conditions. The symbols that catalyze action in human systems are values [Douglas, 1970]. Once ritual has enframed and enhanced [Goffman, 1974] ordered behavior, making it "valuable" [Langer, 1967], emotional triggers stimulate its replication. Ritualized, behaviors like column and line cease to be mere memories; they become moral obligations—and it is moral obligation that writes time into social structures as matter writes time into the rest of nature. That is, once a behavior has become moral, it is rule governed and regularized. Its repetition is catalyzed by values, and by repeating behaviors social structure is preserved. Morality is the storage medium perpetuating the order reached by a society at symmetry breaks.

The patterned behavior that becomes morally valued is, originally, as unpredictable as particular cells in a Benard Instability. No one knows in advance how a Moses or Cincinnatus will react, although historical, post-facto explanations can account for their successes. All that can be honestly said is that, if an action is selected it will be enshrined in morality through ritual. Then ancestral choices provide role models identifying a culture for its entire history. Morality and social structure thus establish a self-referential relationship. Values symbolically encode social structure, while social structure is the context deciphering moral symbols. The mutual representation of morality and social structure makes information "meaningful." In the social context, human beings experience meaning emotionally because of the psychological consequences of enframing. Associating emotions with behaviors, values stimulate the recurrence of preferred behaviors.

Moral organization, the behavioral guidance contained in myths, is the result of a society achieving a symbolic representation of itself, an algorithm mapping the social structure and supplying rules for its replication. Religious myths, as Durkheim [1961] pointed out, were the earliest symbolic representations of a society. They are the first collectively shared "cognitive maps" through which societies describe themselves to their members. Symbolically describing morally sanctioned behaviors, cognitive maps guide the actions relating individual members of a particular society. They are the societal equivalents of DNA, although cognitive maps are not strings of genetic information about biological bodies. The vocabulary in which organizational rules are expressed at the social level has evolved beyond chemistry to ideas and values. But that nature's vocabulary is enriched does not mean its linguistic syntax and grammar changed. Societies replicate when cognitive maps guide behaviors through myths and religions, sacred tales about gods which excite future generations to recreate the past by retelling its glories. To change myths is to defy the gods, while reenacting a myth recreates the world, and the society embedded in it, according to divine plan. Myths, like biological codes, have built-in redundancies that make alterations of their messages unlikely.

Integrating morals into myths and religions, collectively shared cognitive maps make social structures "intentional." But their intention is to perpetuate themselves by replicating successful behaviors. The effects of morals, myths, and religions, moreover, typify the role of bricolage in evolution, for they use the vertebrate capacity for mimicry in new ways. Rather than a single individual mimicking another's acts, through cognitive maps whole societies mimic their collective pasts. That is, through cognitive maps societies organize themselves to collectively preserve the shared experiences of improbable states generated at an instability. Cognitive maps, of course, evolved in concert with the groups embracing them. Eventually, cognitive maps become extremely complex algorithms, for they direct a widening range of specialist behaviors designed to stabilize societies in many different circumstances.

Recording collective experience and communicating group responses meant emerging societies had to map themselves at the system level. The descriptions of positions, roles, and relationships between all individuals in groups driven to improbable states by, e.g., attacks, provided a template for reconstituting these groups in the future. Only a new symbolic language that could be simultaneously present in the minds of every person in a society could meet these requirements. "Consciousness" has finally emerged in its familiar human form—knowing with others and knowing about the self [Fischer, 1978]. Mind, then, literally emerges from matter. but the matter is as much social as biological, for it is in the social structure that knowledge of the civilized world is entrained [Lewin, 1988].

SEX AND SOCIAL EVOLUTION

Organized at a higher level, societies are able to sustain their increased complexity because they can get more work out of individuals. Increased complexity sustained by increased rates of entropy production follows the Second Law. But what made the most primitive peoples work harder and organize more complex societies? Again, the decision need not have been conscious. Rather, working harder may have been a consequence of increased social complexity generated by nonlinearities. The real question is what triggered the positive feedback cycle in the first place.

Many factors no doubt influenced the symmetry breaking leap toward civilization, but I shall single out a necessity for which there seems to be significant biological justification driving the action [Vayda, 1968; Davie, 1929]. Early human groups, the mathematicians tell us, were too small to reproduce themselves [Mann, 1987]. Because of their limited capacity to do work, they had few resources with which to support a large population. Their nomadic behavior meant that population density was minimal. Further, nursing mothers were removed from the breeding pool for long periods. Thus, the number of breeding age males and females was too small to counter the death rate. Any group attempting to reproduce itself endogenously, like thermodynamically closed systems generally, would have been driven back to equilibrium as it died out and disappeared into the environment. Many probably did. Of course, individuals could have reproduced. But, as examples from ethology make clear, group activities are essential to the survival of individuals, once produced—flocks and herds, for instance, warn of predators. Thus, individuals who lived in groups were more likely to survive long enough to reproduce. Survivors, therefore, found ways to breed with neighboring groups, often at the latter's expense.

The incest taboo may originate in the necessity to sustain the social group by expanding the breeding pool. The incest taboo is a good example of a behavior typical of many animals that becomes conscious in social cognitive maps as a moral symbol catalyzing action. It communicates the information that endogenous pairings are unable to sustain the group by converting it into the obligation to breed exogenously. By driving the young males out of the group in search of breeding partners, the incest taboo opened the social structure. Group survival was made possible by accessing necessary environmental resources. But accessing the breeding partners necessary to sustain a group would, often, have been an activity that individuals alone could not perform. Accessing breeding partners for the good of the group required cooperative activity. The reasons were both biological and social.

Since it is possible for one male to replicate in association with several females, breeding age females had priority. Groups that succeeded in acquiring additional female breeding partners survived by generating a positive birth rate. But breeding age females have economic utility to every society. In primitive systems their food gathering and agricultural activities make them the most effective agents for gaining energy. The loss of its females would be a classic example of emerging interdependency, for the results are not only experienced by the abducted women individually. The whole group and everyone in it is jeopardized by their forced departure. It would be reasonable, therefore, for a group to wish to prevent their females being carted off to breed with other males. The group might even resist that happening. Males searching for breeding partners among foreign groups would then have to fight for them [Ortega, 1941], and fighting is an activity that would be much more likely to succeed when done collectively. Of course, raids may have originally been for entirely different purposes, or for no purpose whatsoever. They were probably spontaneous and accidental ''noise'' in the state of nature, especially to the attacked. But when females were brought into a group by raids, a contingent entropy burst had generated new survival information. Repeated a few times, the raiding strategy would become part of the patterned behavior identifying a social structure. With the commitment born of habit, raiding then became a valued behavior sanctified morally.

Still, struggles between two disorganized groups would not be effective. They would simply be another example of the confusion present at near equilibrium conditions. The attacking males, e.g., might lose too many of their own number to use advantageously the few females won. As often as not, disorganized attacks might have been driven off with no captured females. Conflict in these circumstances would have no thermodynamic advantage; it would simply represent the sort of friction that accelerates the degeneration of simple structures to equilibrium.

If, however, one group organized—if it acted in "formation"—it would have a considerable competitive advantage. It would lose less males and gain more females, on the offensive, lose less males and keep more females, on the defensive. The more organized group would then be in position to amplify its emergent structure by decreasing its death rate and increasing its birth rate. That is, the organized group would have a reproductive advantage, and its population would grow to dominance through the natural selection of its social form. Exactly how biological selection takes place is hard to determine, since emergence involves the interaction of dynamic organisms and their dynamic environment. It may be possible to postulate a social process exemplifying the mutuality of effects, however.

STRUCTURE AND SELECTION

Formation raiders immediately, but unintentionally, transformed their environmental niche, the targeted groups. Raiders made their environment the product of human artifice, generating an energy flow with nonlinear consequences that redefined both themselves and their victims. Organized raiding established a gradient driving neighbors farther from equilibrium, where information ordering victimized societies was created. In an improbable state, behavior in victimized groups was correlated. They reached a bifurcation and had to choose between extinction and preserving the information generated by organized raiding. When information was preserved, the selective environment of the raiders changed. It now operated on collective levels, selecting social forms not biological individuals. The raiders themselves were thus forced to intensify their organizational structures.

Warring groups, rather than pristine nature, now constituted the environment in which social structures self-organized and described themselves. Every time raiders attacked another society they defined the victim with increasing specificity. Attacks collapsed a loosely organized group's cloud of evolutionary possibilities, making it more clearly distinct from its environment. The "surround" strategy used in attacks, for instance, encouraged victimized groups to organize a defensive formation called the "circle." Of course, defensive responses had comparable effects on raiders. Improved defenses increased the efficiency of attacked societies, which then released more energy into raiders, feeding back information that stimulated the raiders' further development. Driving attackers progressively farther from equilibrium refined the formation strategy as an adaptive response and forced victims to organize even further.

The raiding cycle thus catalyzed its own nonlinear development. Each iteration sharpened the structures of the involved societies, making a recurrence of the cycle more probable and more effective. Eventually, the cycle acquired a life of its own, subordinating the interests of its components to itself. War, in other words, served as a medium of communication between groups by which each came to decode the other. Mutual interactions reshaped separate groups in each other's images until they began acting together. The result were a true emergent form, one in which each sub-group controlled the behaviors of others and overall behavior regulated each of the sub-groups. Within the emergent society, warriors were made soldiers, victims became slaves, military leaders became chiefs, and warring cities were subsumed into the first states.

New cognitive maps—religions and myths—stabilized efficient forms by symbolizing relations between interacting groups through formalized arrangements of their separate gods. The dominant position in the social hierarchy of the city victorious was matched by the place of its triumphant god in the heavenly pantheon. Mythologizing the relationship between cities made it moral. Thus winners fixed the flow of energy, matter, and information in an efficient pattern through the stabilized cycle, the "civilized" empire. The cognitive maps which appeared with civilized states tracked the flow of information, energy, and matter within emergent systems, perpetuating the functional behaviors necessary to replicate and stabilize that flow.

TECHNOLOGY, COMPLEXITY AND ENTROPY

Following the confusion associated with their emergence, societies stabilized by minimizing the amount of internal entropy produced relative to their level of complexity. They streamlined their internal energy processes. Reorganizing societies meant that a "phase change" occurred, for the relationships between components changed. Changed relations altered behaviors, which amounts to redefining their human components

[Artigiani, 1987]. In emergent structures, new behaviors would be qualitatively different from the past, because an emergent form is a new whole irreducible to either the sum of its parts or to their separate histories. With the emergence of new structures, societies "made up" new people [Hacking, 1985], altering the categories identifying them and providing new models to guide their behavior [Douglas, 1986].

Complex proto-civilized societies emerged as tool using, energy acquiring nomads became weapons wielding, women capturing raiders. Groups that previously had used tools to increase the efficiency of their productive labor now also used weapons and fought for additional resources. It takes energy to produce tools and weapons, entropy is associated with their use, and information is frozen in their shapes. Tools are "energy forms" [Adams, 1988], and the use of new tools guides behaviors in new directions, often triggering a cascade of change. Tools, moreover, are used repeatedly, producing psychological responses similar to those catalyzed by ritual enframing. Tools were actually gods to Homeric Greeks, indicating that the behavior guided by a tool acquires "value." Once incorporated into cognitive maps, values are shared and ensure conformity to collective goals. Societies could then redefine warriors as soldiers. Warriors sought only their own distinction; like Achilles, they were unpredictable. As Shaka Zulu proved, however, soldiers can be trained by rule-bound procedures to accept a discipline that sacrifices their whimsical individual needs and allows societies to pursue long-range goals. Once civilized, therefore, war becomes "rational;" through it societies subordinate subjective personal intentions to economic and political ends. Rationalization is an example of increased social interdependency, of wholes becoming greater than the sums of their parts.

Weapons, like all other technologies, preserve information-guiding behavior in concrete form. To learn the use of weapons was to make possible the role of soldier and redefine the human components of social systems. Weapons, too, could make the acquisition of environmental resources more efficient, increasing the flow of energy through social systems. More energy makes more complex societies possible, and weapon-wielding soldiers emerged as specialists within the social hierarchy. Their appearance, however, had the expected increased entropy price, for, in addition to the death they deal, soldiers consume wealth during their years of drill and training. Other specialists, slaves and craftsmen, were needed to produce the soldiers' food and necessities.

Specialization made societies more complex, and increased complexity led to the growth of managerial elites, which were apparently needed to supervise the flow of resources through societies. Through the work of managerial elites guided by the blueprints of collective cognitive maps, the structure of a far-from-equilibrium society was preserved. Thus the society became the effective processor of information, energy, and matter from an environment dominated by other societies. Organized, stable social institutions think [Douglas, 1986] through the actions triggered by communicating ideas and values between individual humans. Preserving the flow of information, energy, and matter into and through the newly defined structure meant that formation was extended and became permanent, giving rise to the state.

The state is an organizational network designed to administer society. With the emergence of the state, society had further developed by specializing behaviors among its component parts [Bonner, 1988]. Specialized component behaviors, in turn, led to a greater need for organization. It was no longer enough for raiding parties to be formed by a command hierarchy; to minimize internal entropy it was now necessary for the command hierarchy to extend its ordering strategy to the behaviors and relationships of the entire population, harnessing the cooperative behavior of all [Deleuze and Guattari, 1987]. Accessing greater energy, diversifying the population, introducing hierarchy, and producing external entropy at increased rates, society had irreversibly evolved farther from equilibrium. Since the group activity permitting that move was war, it is not surprising that the form taken by cooperativity in civilized societies was coerced [Schmookler, 1984]. Coercion, of course, was ensured through the more rigorous organizing agencies, religion, writing, and slavery, introduced by the command hierarchy.

THE EMERGENCE OF HUMAN CONSCIOUSNESS

With the emergence of proto-civilized societies, nature—in the form of organized societies—had evolved so far-from-equilibrium that human behavior was systemically orchestrated. When groups with formation succeeded, they had become more complex by organizing humans who, in turn, were expected to carry out more varied actions on behalf of their groups. Groups acquired collective identities as reproducing populations, while specialization increased dependency and the awareness that human components were members of entities greater than themselves. The success of organized groups, of course, called attention to them, making their human components more aware of themselves as they were made aware of the group with which they shared knowledge.

Sociologists indicate that conflictual and consensual validation are two keys to human identity [Stein et al, 1960]. Conflictual validation is negative; it tells people they are different from those with whom they fight. Anti-semitism, nationalism, and racism are the most common forms of conflictual validation. Consensual validation is positive. It tells people they are accepted by those making up a particular group. Military "esprit de corps" cultivates consensual validation. Consensual and conflictual validation encouraged identity, the hallmark of human consciousness. Its emergence represented new information created through the nonequilibrium conditions affecting social relations.

Conflictual validation generated by formation raiding reflected back on the group to create consensual validation. In the same way the cocoon made by a larva changes the organism into a butterfly [Prigogine and Stengers, 1977], the raiding information ordering a society created the defensive boundary conditions redefining it. Redefined, human societies described themselves algorithmically by new "cognitive maps." The new cognitive maps described highly particularized collective experiences in symbols identifying unique social structures. Societies were now readily distinguishable from their environments and from each other, while the members of each society occupied classes that were similarly distinct.

The earliest societies, influenced by feedback from the physical environment, had organized themselves in the image of the natural world. Like all natural structures, they were statements about—"mappings" of—the environment in which they were embedded [Bateson, 1972]. The societies emerging in the river valleys were more successful hypotheses, but they, too, mapped environmental reality. However the new social structures were collective statements made by organized human beings, whose cooperative work produced the environmental realities mapped by the new mythologies. Their environment included the legacy of their collective pasts, whose symmetry had been irreversibly broken by conflict and permanent settlement. In the new environment, humanity encountered itself, and collective hypotheses about the world included social as well as physical nature. Thus, in addition to statements about natural reality, the new mythologies made statements about the collectives making the statements. The emergent river valley societies were aware of their own existence; their maps of the world included themselves, as their "origin" myths make clear.

Integrating and specializing a large number of people in the work of collectively mapping societies and their environments created the conditions for human consciousness to emerge. Coordination required monitoring the status of society, evaluating environmental dynamics, and adjusting responses to meet projected needs. Monitoring, evaluating, and adjusting are mental activities which, once symbolic language existed, were carried on collectively. That is, through symbolic language encoded in cognitive maps, the experience of any individual became the property of all [Chatwin, 1988]. Meanwhile, through raiding and other environmental fluctuations, the information being communicated was equally collective. Information collected when individuals monitored the environment rippled through entire societies, forcing all to adjust. When people become mutually dependent on symbolic information, society had literally become a whole greater than the sum of its parts. The information communicated to any individual—and shaping his or her awareness—was based on collective experience.

Monitoring, evaluating, and adjusting the state of societies in their environments are probably examples of "parallel distributive processing," for, through shared cognitive maps, individuals now used one another's brains and society used the brains of all. The connections and dependencies between people meant that any individual's decisions were, at least partly, the results of everyone else's thoughts. The information in any person's brain, therefore, extended far beyond personal experience, and much of that information was from a perspective that transcended the individual communicating it. Thus consciousness, in the first instance, is an evolved consequence of organized group behavior. It is "mind," a social phenomenon [Fischer, 1989] that is an emergent characteristic [Hillis, 1988]. Consciousness is society thinking itself.

But consciousness is as much the individual's awareness of him or her self as it is the collective awareness of a society. The two meanings, thinking about the self and thinking with others, emerge together. Organizational structure made the society aware of its environment collectively. Stabilizing a far-from-equilibrium structure in a dynamic environment forced individuals to adjust to the conflictual experiences of their society. Each adjustment would be a learning experience in which new information entered the individual's brain. The shifts from state to state caused by the influx of information, each slightly changing the relations identifying a person, made the individual self-aware [Pribram, 1980]. Repeated iterations of a societal form focused the sense of consciousness until it simultaneously became the basis for the identity of groups and for identifying individuals. Sharing knowledge with others about the society in which all participated led eventually to increased consensual knowledge of the individual selves constituting the society.

Consciousness, in its social sense, is the mutual sharing of coded information orchestrating interdependent action toward the collectively sanctioned goals established by recorded societal experience. Consciousness emerged as a direct consequence of the nonlinear processes by which social complexity increased. Driven by the intense violence fluctuating structures within the boundaries of the river valleys—or other exchanges of energy, matter, and information—societies were so far-from-equilibrium that even the slightest perturbation could trigger a cascade of events collapsing whole structures. The balancing act had become too precarious, therefore, to allow events to go unnoticed by the command hierarchy. The administrative network regulating society had to be informed, alert, and quick to respond. Only aware individuals correlating their behaviors through a common code of communication could provide the flexibility necessary.

COMMUNICATION

The first examples of column and line required no such sophisticated organization. They were merely improbable states measuring the departure of early raiding groups from their hypothetical near-to-equilibrium origins. Once in existence, however, column and line were positionings of the members of a raiding society that proved advantageous in the struggle for resources. Thereafter the distribution of members of successful raiding groups were not equiprobable. At least in certain circumstances, they were more likely to be in the positions constituting column and line than in any other possible state.

Column and line indicated that when raider x followed y and z, an attack was successful. A way to store this newly discovered information, and communicate its message to succeeding generations, was needed. The simplest way to store information is to repeat the message communicating it over and over again. Campbell [1982] calls this primitive device for storing information "context free redundancy." Context free redundancy communicates "meaningless" information. It only stores information describing an improbable state. It does not describe the relationships between elements in that state, since repetition alone can refer only to so simple a state that the components constituting it are virtually independent of each other. Context free redundancy ensures accuracy of communication, but it restricts the messages that can be sent. It leads raiders to insist that z always precede y and x. In effect, context free redundancy means a source has only one message to send.

A simple structure functioning in a narrowly prescribed environment needs to know very little about the world but has to preserve that knowledge exactly. If, however, a structure operates in a broader or richer environment, where changing conditions are regularly experienced, then it must store more information. Every aspect of its enriched environment must be described, and, if these aspects change, the structure must store information about each of its several accessible states. Repeating the same message over and over becomes counter-productive, for there will be environments which the message—i.e., the positioning of members of the group—does not describe. Context free redundancy, for instance, would not permit others to raid when x, y, and z were fishing. More than one message must be sent in expanded and enriched environments, where a structure must store greater information about the world.

If the environment in which a structure is embedded is not only broad or rich enough to offer various states but dynamic enough for the states to succeed themselves unpredictably, then the structure must be prepared to describe unexpected or wholly new environments as well as select between messages. Gatlin [1972] points out that any structure which had locked itself into the eternal repetition of a single message could never adapt to the rigours of an enriched, expanded environment, whose dynamics are such that any inaccuracy in the storage or communication of structural information is increasingly costly. A society which only knew how to organize an attacking column, for instance, would not know how to organize a circle for defense. To survive in changing conditions, a structure must be able to store a vast amount of information and communicate different messages quickly. At this point, it is not the improbable states in which the elements of a structure are arranged that assures survival but the relationships between the elements. Evolution now depends on the ability of a structure to shift between states as environmental conditions vary. To store information regarding a variety of states and to smooth transitions between them requires a new form of redundancy.

This more sophisticated form of redundancy is sometimes called Shannon's Second Theorem [Shannon and Weaver, 1949; Shannon, 1951]. It postulates that message accuracy could be ensured by using codes whose internal constraints govern the relations of symbols. Second Theorem codes can communicate more information safely. Such codes reflect situations in which interdependency gives meaning to the behavior of individuals even in altered circumstances. Interdependency creates a context in which components operate. Second Theorem codes map that context; they encode the rules societies use for mapping environments by simulating their worlds in their structures. Rules reduce the number of possibilities in an informational sequence, for now, e.g., the position of any raider depends on his contextual relationship with the others. The code records information about both the relationships and positions of components. New generations will not wander the landscape haphazardly. Obliged to organize themselves into columns, lines, and circles, successive generations will preserve social order amidst environmental fluctuations.

According to Gatlin, beyond a certain barrier, where the informational flow becomes too dense to prevent message disruption by simple repetition, evolution depends on switching to this new form of redundancy. The cross-catalytic processes by which raiders and their victims exchanged energy, matter, and information carried social systems to this degree of complexity. Raised to new levels of complexity and reshaped in one another's image, conflicting groups changed the ways in which message accuracy was secured. They created cognitive maps as their codes. Campbell calls this new form of redundancy "context sensitive."

Second Theorem redundancy emerges in human societies when they become conscious of themselves. Information is then stored as concepts and communicated symbolically. When a group of raiders, for instance, forms the concepts "column" and "line" they no longer need to store that information by positioning their bodies. They can move beyond non-verbal media like hand-holding, rituals, and dances, which stored only limited information through context free redundancy. Once societies become conscious by constructing cognitive maps that correlate the actions of individuals, information can be preserved by symbolized values in human minds. A column or line is no longer a particular ordering of specific individuals x, y, and z describing a single state. Symbolized as values catalyzing action, the cognitive maps produced by raiding societies described the relationships between any individuals organized in moral contexts. Cognitive maps prescribe functional behaviors allowing societies to process information from their collective environments.

THE ACCELERATION OF CULTURAL EVOLUTION

One frequently voiced criticism of Darwinian evolution is that it does not explain why evolutionary survivors do survive. The model offered here helps resolve that problem, for societies armed with efficient and flexible codes for storing and communicating information were able to survive even in changing environments. Through representations of interdependency, societies mapped both themselves and their world. Unlike the representations of material nature, conscious societies were not limited to describing external environments. By representing themselves in their cognitive maps they could plot the effects of their actions on the world. Thus, societies learn how to learn; they can plan techniques for surmounting obstacles in advance [Schull, 1988].

By representing relationships, symbolic information allows the men in a column to perform a maneuver, transforming their structure into a line for attack when they reach their target, for instance. Context sensitive redundancy thus makes evolution economical at the level of collective reality. It provides structures with a meta-rule, a rule for the making of rules appropriate to changing circumstances. Cognitive maps can direct specialist behaviors, communicate the information they generate, and coordinate its consequences. Societies adapt because components can be used for purposes varying with circumstances. The lower classes of a society, e.g., can be made to dredge irrigation ditches, plow fields, harvest crops, erect public buildings, or construct fortifications, depending on the environmental needs of the moment. Their position in society does not change, but the different kinds of work they do changes and those changes affect every one else in the society.

The advantage of depending on more efficient codes to ensure that information was stored and communicated accurately accelerated the rate of cultural evolution. Environmental selection and random mutation were not the only engines driving evolution, for now societies could self-organize by manipulating the symbols representing their environments. Since ideas and values, the components of cognitive maps, are cheaper than genes, they could be treated *as* reality and societies could imaginatively explore ever more effective configurations. A social structure did not have to explore its environment by actually organizing a life and death material test of an evolutionary hypothesis. Possibilities could be explored conceptually, and conceptual explorations are even more efficient at testing an environment than play and ritual dance.

Ideas and values permit social systems to anticipate and explore future developments [Rosen, 1985], and, after social structures included models of themselves in their environmental maps, that is precisely what they were able to do. Armed with mental images of column and line, for instance, societies could focus on the relationships defining them, manipulating and adapting column and line in advance as other societies in their environment became more efficient in attack and defense. But symbolically encoded mental images linked the components of societies interdependently, allowing the collective to respond as a whole to environmental perturbations.

CONSCIOUSNESS AND HOMEOSTASIS

Real societies interact in a variety of ways to exchange energy, information, and matter. For the sake of simplicity, I have based my model of the process of societal self-organization at increasing levels of complexity on violence. Violence is a relatively simple activity, whose repetition can easily generate extremely complex patterns of order. Iterations of commercial, intellectual, and agricultural activities were, no doubt, as important, if not more so. But they are too subtle to analyze briefly. Still, if it is a mistake to conclude that violence alone made societies more complex, it seems equally clear that the form of social complexity called "civilization" usually appeared in an environment of social violence [Carniero, 1970]. Civilization emerged in river valleys, where settled groups with vested collective interests could not flee the violence of one another or the nomads on their peripheries. Particular boundary conditions and the intensity of environmental fluctuations drove groups through ever greater violence until they stabilized as civilized empires.

Thus war, like everything else, is an historical construct [Mead, 1940]. It is not a fundamental given but an embedded structure founded in a random response to a contingency and based upon specific circumstances and actions. Nevertheless, in an historical nature, the legacies of the past are never wholly forgotten. Therefore, even the societies which escaped the feedback cycle leading to civilization were affected by violence. Their conditions, however, allowed them to stabilize less far-from-equilibrium, where their emergent consciousness could regulate violence at more moderate levels. Amerindians, for instance, often made war a sport, giving priority to collecting "coups" rather than killing one another [Farb, 1978]. In other contexts, ritualized bride rape preserved convention, even though new, non-violent means of exogenous marriage had been arranged.

In the model offered here, war originated as a system response to negative population growth. But exogenous breeding practices, at least among surviving social structures, solved that problem all too well. In the absence of the resources present in the river valleys, population growth became excessive. Environmental pressures now required war as an institution to reverse its role. But Marvin Harris [1974] argues war was incapable of curbing the population growth it released. In locations where boundary conditions and resources did not lead civilization to emerge, too many negative control mechanisms had been built in for war, alone, to effectively check population growth. Even when real wars were fought, they almost always stopped long before a population was decimated. But to wage war at all requires surplus males. Maintaining enough males to wage war encourages female infanticide, making war, at best, an indirect means of population control. Since males are largely reproductively redundant, controlling the number of females reaching sexual maturity effectively balances population to resources. Societies now only needed to waste the number of excess males, and war is adequate for that.

The rhythm of war and peace Harris describes in New Guinea exemplifies societal level regulation of the cycle of population growth and environmental depletion, with appropriate responses guided by societal cognitive maps. Critical resources, in this case, pigs, are symbolic indicators of excess population, signaling wars should be fought to preserve the homeostatic balance between population and environment. The pig population, valued because of its contextual position in mythical cognitive maps, tracks human environmental demands. When the pig population reaches a critical threshold, specific behaviors are triggered. Pigs are slaughtered and wars declared. When war produces sufficient casualties, the loser withdraws from both the field and its home territories, providing an ecological respite. Years of peace follow while pigs and people multiply—and tales are told of the outrages perpetrated by the enemy, preparing the way to renew the cycle in future generations. Thus through the mythologically controlled rhythm of war and peace, the societies Harris described use their Second Theorem codes to think themselves into descriptions of altering environmental states.

ORGANIZATION, CONSCIOUSNESS, AND AUTONOMY

The emergence of consciousness, collectively and individually, reveals the revolutionary nature of explanations inspired by the new scientific paradigm [Roque, 1988]. A linear causal relationship can no longer be expected [Roque, 1985]. Organizing human groups farther from equilibrium did not "cause" consciousness. Instead, a reflexive relationship between components and levels of structures exists [Wohlmuth, 1988]. As the several elements interact they are mutually selecting and selected, each defining the other in terms congenial to the emerging whole. Thus, increased complexity on the collective level is as much dependent on individual humans who are more conscious as it is the cause of their consciousness. To survive far-from-equilibrium a structure must have constant information about its embedded environment and make continuous adjustments to its fluctuations. Conscious individuals perform these monitoring and governmental functions. People and their societies co-evolved in a coupled relationship, the progress of each calling forth progress in the other. The tragedy of civilization is that the conscious autonomy of the few was purchased at the expense of the enslaved drudgery of the many [Boulding, 1964].

Ancient Greek military organization exemplifies the way in which increased complexity makes both the society and privileged individuals more conscious. The Greeks organized their armies around an infantry unit called the "phalanx." The phalanx was a solid square composed of lines of armored soldiers, "hoplites," each of whom wielded a stabbing spear. A phalanx fought as a unit. Drilled, disciplined, and determined, its members ran into battle shoulder to shoulder. Each man was free to fight because his neighbor's shield protected him. If all coordinated their efforts, tremendous energies were released. But if any one faltered—lowered his shield, turned to run, failed to step into the gap created when a comrade dropped, or even tripped on a rock—the whole unit could disintegrate in a moment. The phalanx line was a chain, literally as strong as its weakest link. It possessed a high degree of interdependency.

To strengthen each link, the Greeks invented democracy, for only soldiers persuaded of the rightness, profitability, or necessity of their cause could stand the terror of shock combat [Artigiani, 1985]. Each hoplite had to be consciously committed to the collective and its chosen acts [O'Connell, 1989]. Socrates, the first philosopher to teach that "the unexamined life is not worth living," was, after all, a hoplite member of the Athenian phalanx; and Athenian consciousness, expressed as something approximating "nationalism," was one cause of the endemic wars that ravaged Greece. Thus by organized cooperativity both the individual and the society acquired a greater sense of their own identities—while ever more destructive warfare measures the greater entropy produced by these more complex societies.

The paradoxical relationship between component autonomy and structural organization can also be resolved when the far-from-equilibrium paradigm is applied [Pattee, 1973]. Roman military history provides examples [Culham, 1988]. Roman military units were tightly organized as "cohorts." The cohorts were joined together to form armies, which were structural copies of themselves on a larger scale. In battle, however, commanders of small components, the "maniples," were permitted considerable autonomy because the structure over-all was tightly organized. Thus, individual commanders could respond autonomously to local conditions, their highly disciplined units performing unplanned wheeling or flanking movements at bifurcation points that often drove battles to stable states favorable to the Romans. Such maneuvers were only possible because Roman armies were thoroughly disciplined and clearly organized, making choices possible and meaningful. Organization and autonomy, like entropy and information, are complementary.

THE COEVOLUTION OF LANGUAGE AND SOCIETY

Traditional linear causality implies symbolic language made possible the evolution of social complexity. But, of course, without social complexity there would be no need for symbolic language. That is, in an evolutionary world where everything depends on everything else, neither is conceivable without the other. The two must have co-evolved as parts of an emergent level of reality. It follows that the history of language tracks the evolution of society, offering a kind of test of the validity of the model developed here.

Little is known about the origins of symbolic language, since no historical record exists. But linguistic analyses, philosophical speculations, and anthropological studies indicate a process in which physical, self-explanatory gestures evolved into conventionalized symbolic sounds capable of communicating increased information. Integrated languages emerged from aggregates of separate sounds in the same way that civilized societies self-organized from primitive groups: Language was part of the process by which behaviors were correlated and information recorded far-from-equilibrium. By relating thermodynamic and informational entropy, the notion that contemporary natural science can be applied to the humanities is supported.

The first language was probably gestural, as Condillac assumed [Condillac, 1746]. Gestures were adequate to communicate in simple groups, for the earliest gestures were self-explanatory statements about the particular, immediate desires of people in direct contact with one another. But when more complex signals needed to be sent describing, say, formational changes from column to line during a night raid, a verbal medium was needed.

The first words must have been randomly chosen, since a century of linguistic analysis has failed to link sounds and the objects they conventionally represent. Moreover, Cassirer [1945] argues that the first named objects must have virtually overwhelmed the minds of the first namers. To name, Goethe pointed out, is to create, and the shock of recognition following on the sudden differentiation of an object from its environment must have monopolized the mind of the person perceiving it. Since overwhelmed minds cannot function long, the first names were transient as well as individualized. Like the primal groups in which they were created, the first named entities were highly entropic, fragmented, and unstable.

There must have been an infinity of sounds, for instance, that expressed an individual's alarm upon seeing a raiding formation poised for attack. The one chosen was spontaneous and unpredictable. Moreover, since other members of the victimized group would have expressed their fears individually, many sounds no doubt were made. The stochasticity of selected sounds, on the one hand, and the entropy of equiprobable competing noises, on the other, meant that each individual was using a biological capacity to make noises but no language existed. Producing sounds is one thing; perceiving their meaning is another [Lieberman, 1975]. Until both potentialities were realized, language did not exist.

To make language possible, a public "shared experience" [Wells, 1987: p. 119] had to evolve. Since words are not self-explanatory but arbitrary conventions, a sound in itself communicates very little information. A cry of alarm could have meant any number of things, from "falling tree" through "wild beast" to "raid." But if an individual happened to repeat the same cry especially loudly when formation raids occurred, then the shared experience of victimization would establish a context in which that particular noise was decoded. It would have an experiential meaning to a raided society. Thereafter, individuals could communicate over macroscopic distances, correlating their behaviors. Each would know what is happening to all, and even though nothing yet may have happened to him or her individually each would react cooperatively.

At the same time that they were shouting fearfully, of course, attacked individuals would have been scurrying randomly in every direction. But if the raiders attacked in a "surround," then the attack itself would have herded the victims into a "circle" for defense. Repeated use of a cry conventionally deciphered as "attack" and followed by the forming of a circle would have reduced both the thermodynamic entropy—the physical energy wasted in random flight—and the informational entropy—the chaos of conflicting noises. Thus, as the gradient of successive attacks pushed a group far-from-equilibrium the behaviors of its people were correlated, and their correlated behaviors were recorded in an elementary language as crude as the first raiding formations. But in its emerging language, a group would have found the way to preserve behavioral information giving them, collectively, a selected structural advantage. Language gave the group "logical depth," for they could through a "word" replicate the organization learned through bitter, costly experience. But the word did not create the structure any more than the structure created the word. Both emerged together in an experience that is only explicable "top down," from the perspective of the self-organized social whole.

Linguistically preserved defensive information represents problem solving on a collective level, for there is nothing aside from flight an individual can do to protect against formation raiding. Random motions, as random cries, are context free, "meaningless" information. By organizing, the society solves problems beyond the individual's capacity. Then the society has learned to process information collectively. But effective information processing requires a program, an arrangement of symbols that allows for the in-put of environmental information and the production of structural information. The deification through naming of critical environmental realities provided the symbols; their arrangement in cognitive maps provided the programs. Cognitive maps representing the society in its world created the symbol decoding context making information meaningful on the linguistic level.

Because the perception or decoding of sounds is context sensitive, and because the context reflects the unique experiences of a particular group, every language has a "personality" [Jennings, 1984]. The existence of a set of context sensitive conventional sounds decipherable only by members of a particular group, in turn, encouraged conflictual and consensual validation of the group as an organized entity clearly differentiable from other societies in its environment. More clearly identifying the group linguistically helped distinguish its organizational structure, for the ever more refined meanings given to arbitrary oral conventions meant the group was streamlining its self-image by reducing the informational entropy in its language. Sharpened linguistic meanings meant words could be used to improve organizational patterns, increasing thermodynamic efficiency. Co-evolving thermodynamic and informational processes interacted in a nonlinear positive feedback loop increasing the complexity of each.

Once sounds achieved a public meaning, words stabilized, acquiring an existence of their own. As collectively shared records of social experience, words endured. They could be studied, and their vague, multiple meanings reduced to single, ever clearer definitions. Archaic deities record this entropy reducing process, for successful gods had a history remembered in their multiple names. Coming from aggregated sources, the deities reflected disconnected experience. But as group identity sharpened, gods became associated with particular societies and particular identifying experiences. As their names became singular, they reported on the emerging identities of the peoples with whom they were historically associated.

In the public domain words also multiplied, permitting more specific environmental information to be communicated. The more environmental information communicated, the farther-from-equilibrium a society could move and the more effective its homeostatic choices became. But to communicate ever denser messages the information symbolized by words had to be formally patterned. The grammars which developed followed pathways as diverse and unique as vocabularies. However, grammars, as B. L. Whorf [1956] pointed out, constrain not just the messages sent but the kinds of thoughts peoples have. As minds self-organized between brains trained to distinguish certain sounds arranged in specific ways, they moved away from the Babel of chaotic communications toward ever more efficient media, whose hierarchical organization reflected "deep structure" [Chomsky, 1957]. Again, the appearance of a publically accepted grammar allows it to endure, be reflected upon, and be refined. Refinements mean a grammar, like other Second Theorem Codes, would be able to reduce internal entropy, communicating more information with reduced noise as societies developed more complex structures in dynamic environments laced with man-made "energy triggers."

But informational structures do not appear haphazard. They have very particular organizational rules, and those rules have a real world basis. It is, of course, true that people can invent any sort of grammar they wish, provided only that the possibilities present in the brain include them. But a public language reflecting shared experiences cannot communicate randomly. Whatever the stochastic foundations of a particular set of sounds, once publically endorsed that set has been environmentally selected. It will be socially replicated. So, too, will be the grammars organizing words, the rules ordering the communication of conventional oral symbols. The environment that selects for sounds is the shared experiences of an organized society.

Perhaps the environment selecting for grammars is the organizational structure of the society. That is, the relationship between named entities perceived as meaningful maps the relationships between people in a society. If so, then the thermodynamic fluctuations shaping a social structure will be recorded in the grammar making possible the encoding of experience. As a society becomes more efficient by increasing the connectivities between its members, its language becomes more effective by formalizing grammatical rules. But the language is recording the information structuring the society. If the two structures are mutually selecting, then the emergence of a new level of reality, the social reality of human consciousness, has been mapped.

The introduction of writing follows the same evolutionary path and tracks the same thermodynamic realities. Writing measured the distance a society traveled from equilibrium. In preliterate societies, aurality permits unconscious, incremental alterations in social structure, for there is no established standard against which to test the validity of any contemporary recitation of tradition [Goody, 1977; Detienne, 1986]. Communicated orally, preliterate traditions vary subtley with changing circumstances. Inflections alter, stanzas are omitted or shifted, and words are changed, depending on present circumstances. In the absence of a written record, slight changes go undetected.

Moreover, incremental variations in oral traditions would be invisible to a primitive group because the actual version sung or recited would track contemporary environmental realities. Thus, preliterate groups adopt unconsciously to their environments, with which they are in a nearly equilibrium relationship. Literate societies, however, recognize any change in their recorded cognitive maps, and negative feedback mechanisms act to punish transgressions of the written word's authority. The social world, of course, is then less stable. Mismatches between cognitive map and actual experience soon become apparent.

When a society cannot unconsciously change its cognitive map to retain equilibrium with its environment it acts to change the environment. This is one way of recognizing far-from-equilibrium conditions. The society is aware of itself as an organized entity distinguishable from its world. The sort of intentional behavior that acts upon an environment to stabilize a society further testifies to its complexity; in fact, historical convention defines "civilization" as the creation of a society that dominates its environment rather than vice versa [Toynbee, 1961]. Obviously, more environmental information must be collected, stored, and communicated in far-from-equilibrium conditions. Writing is the medium making that increase in informational density possible. But again, the emergence of writing, as well as its own self-organization by the reduction of internal entropy as scripts became alphabets, tracks a social reality comprehensible in terms of irreversible thermodynamics.

Thus, language and society can be seen as co-evolving in a mutually reinforcing process making possible self-organization at increasing levels of social complexity. Without decoding, experience cannot be shared. Without collective experience, there would be no information to share. And without energy fluctuations collective experiences would have neither emerged nor evolved. The model offered here, like any scientific theory, does not account for every step in this process. Nor did it attempt to empirically validate itself. But this brief overview would seem to indicate that a convergence of the natural and human sciences is feasible. At least conceptually, the model derived from quantum randomness, thermodynamic temporality, and self-organization appears friendly to social theory.

Several colleagues read early drafts of this manuscript or graciously made suggestions concerning it. Alicia Roque, particularly, made many creative contributions. Ervin Laszlo, Stan Salthe, R. Ulanowicz, Jon Schull, William B. Haskett, and Bob Crosby were also helpful, and I wish to thank them especially. The arguments presented here are intended to follow those of Ilya Prigogine, although neither he nor those who offered advice should be held responsible for errors, misconceptions, or speculations.

REFERENCES

J. Alcock [1979] *Animal Behavior.* Sunderland, Mass: Sunderland Assocs.

P. M. Allen and J. M. McGlade [1987] "Evolutionary Drive: The Effect of Microscopic Diversity, Error-Making, And Noise," Forthcoming

R. Artigiani [1985] "Language, Logic, And Energy," in I. Progogine and M. Sanglier (eds.) *Laws Of Nature And Human Conduct.* Brussels: Task Force Of Research Information And Study On Science

———— [1987] "Revolution and Evolution: Applying Prigogine's Dissipative Structures Model," *Journal of Social and Biological Structures.* 10: 249–64

H. Atlan [1985] "Natural Complexity And The Self-Organization Of Meaning," in *The Science And Praxis Of Complexity.* Tokyo: UN University

G. Bateson [1972] *Steps To An Ecology Of Mind.* New York: Ballantine

C. H. Bennett and R. Landauer [1985] "The Fundamental Physical Limits Of Computation," *Scientific American* 253

J. T. Bonner [1988] *The Evolution Of Complexity By Means Of Natural Selection.* Princeton: Princeton University

K. Boulding [1964] *The Meaning Of The Twentieth Century.* New York: Harper

J. Bronowski [1973] *The Ascent Of Man.* Boston: Little Brown

R. Callois [1961] *Man, Play, And Games.* Chicago: Free Press

Jeremy Campbell [1982] *Grammatical Man.* New York: Simon and Schuster

Joseph Campbell [1983] *The Way Of The Animal Powers.* San Francisco: Harper and Row

R. L. Carniero [1970] "A Theory Of The State," *Science,* vol. 169, pp. 733–38.

E. Cassirer [1946] *Language And Myth.* New York: Dover

E. Chaisson [1987] *The Life Era.* Boston: Atlantic Monthly

B. Chatwin [1988] *The Songlines.* New York: Penguin

N. Chomsky [1957] *Syntactic Structures.* New York: Mouton

E. Colson [1952] "Social Control And Vengence In Plateau Tonga Society," *Africa,* XXIII

E. B. de Condillac [1746] *De l'origin et du progress du language.* Paris

J. Corliss [1988] "The Dynamics of Creation: The Emergence of Living Systems in Archaen Submarine Hot Springs," Forthcoming

R. Coulborn [1969] *The Origin Of Civilized Societies.* Princeton: Princeton University Press

Social Evolution 127

V. Csanyi [1987] "The Replicative Evolutionary Model Of Animal And Human Minds,"
    World Futures, vol. 27, no. 3, pp. 161–202
P. Culham [1987] "Choices, Constraints, And Chaos: Error And Failure In Ancient Military Engagements," Forthcoming
M. R. Davie [1929] The Evolution Of War. New Haven: Yale University Press
P. Davies [1988] The Cosmic Blueprint. New York: Simon and Schuster
G. Deleuze and F. Guattari [1987] A Thousand Plateaus. Minneapolis: University of Minnesota
M. Detienne [1986] The Creation Of Mythology. Chicago: University of Chicago
N. Dixon [1976] On The Psychology Of Military Incompetence. New York: Basic Books
W. G. Doty [1986] Mythography: The Study Of Myths And Rituals. University, Alabama: University of Alabama
M. Douglas [1986] How Institutions Think. Cambridge: Cambridge University Press
——— [1970] Natural Symbols. New York: Pantheon
W. H. Durham [1976] "Resource Competition And Human Agression: A Review Of Primitive War," The Quarterly Review Of Billogy, 51:385–415
E. Durkheim [1961] The Elementary Forms Of The Religious Life. New York: Colliers
M. Eliade [1975] Myth And Reality. New York: Harper Colophon
P. Farb [1978] Man's Rise To Civilization: The Cultural Ascent Of The Indians Of North America. New York: Bantam
A. Ferrill [1985] The Origins Of War. New York: Thames and Hudson
R. Fischer [1977–78] "On Dissipative Structures In Both Physical And Information Space," Journal Of Altered States Of Consciousness, vol. 3, no. 1, pp. 61–68.
——— [1989] "Why The Mind Is Not In The Head," forthcoming in Diogenes
M. Fortes and E. E. Evans-Pritchard [1961] African Political Systems. New York: Oxford
M. Foucault [1978] The History Of Sexuality. New York: Random House
H. Frankfort [1948] Kingship And The Gods. Cambridge: Cambridge University
——— [1973] Before Philosophy. Baltimore: Penguin
L. L. Gatlin [1972] Information Theory And The Living System. New York: Columbia
P. Glansdorf and I. Prigogine [1971] Thermodynamic Theory Of Structure, Stability, And Fluctuation. New York: Wiley
E. Goffman [1974] Frame Analysis. New York: Harper
J. Goody [1977] The Domestication Of The Primitive Mind. Cambridge: Cambridge University
I. Hacking [1985] Restructuring Individualism. Stanford: Stanford University Press
M. Harris [1974] Cows, Pigs, Wars, And Witches. New York: Vintage Books
J. Harrison [1924] Mythology. New York: Harcourt
W. D. Hillis [1988] "Intelligence As An Emergent Behavior; or, The Songs Of Eden," Daedalus, vol. 117, no. 1, pp. 175–190
C. S. Holling [1988] "Preserving The Sources Of Renewal," in R. Artigiani (ed.) Technological Innovation And Institutional Adaptation. Annapolis: USNA
——— [1986] "The Resilience of Terrestrial Ecosystems: Local Surprise And Global Change," in W. C. Clark and R. E. Munn (eds.) Sustainable Development Of The Biosphere. Laxenburg: IIASA
J. Jaynes [1976] The Origin Of Consciousness In The Breakdown Of The Bicameral Mind. Boston: Houghton-Mifflin
F. Jacob [1982] The Possible And The Actual. New York: Pantheon
G. Jennings [1984] World Of Words. New York: Atheneum
S. Kauffman [1984] "Emergent Properties In Random Complex Automata," Physica D, pp. 145–156

S. Langer [1967] *Mind: An Essay On Human Feeling.* Baltimore: Johns Hopkins
E. Laszlo [1987] *Evolution: The Grand Synthesis.* New York: Simon and Schuster
R. Lewin [1988] *In The Age Of Mankind.* Washington: Smithsonian
P. Lieberman [1975] *On The Origins Of Language.* Lanham: University Press of America
A. MacIntyre [1988] *Whose Justice? Whose Reationality?* South Bend: University of Notre Dame
M. Mann [1987] *The Sources Of Social Power.* Cambridge: Cambridge University Press
W. H. McNeill [1963] *The Rise Of The West.* Chicago: University of Chicago
M. Mead [1940] "Warfare Is Only An Invented—Not A Biological Need," *Asia,* XL
B. J. Meggers [1954] "Environmental Limits On The Development Of Culture," *American Anthropologist,* 56: 801–824
G. Nicolis and I. Prigogine [1977] *Self-Organization In Non-Equilibrium Systems.* New York: Wiley
R. O'Connell [1989] *Of Arms And Men.* Oxford: Oxford University
J. Ortega y Gasset [1941] *History As A System.* New York. W. W. Norton
H. Pattee [1973] *Hierarchy Theory: The Challenge Of Complex Systems.* New York: Braziller
K. R. Popper [1982] *Quantum Thoery And The Schism In Physics.* Totowa, N.J.: Rowan and Allanheld
K. H. Pribram [1980] "Mind, Brain, And Consciousness" in J. M. and P. J. Davidson (eds.) *The Psychobiology Of Consciousness.* New York: Plenum
I. Prigogine (1980) *From Being To Becoming.* San Francisco: Freeman
——— [1982] "Man's New Dialogue With Nature," Preprint
——— and I. Stengers [1977] "The New Alliance," *Scientia,* vol. 112
——— ——— (1984) *Order Out Of Chaos.* New York: Bantam
A. Rae [1986] *Quantum Physics: Illusion or Reality?* Cambridge: Cambridge University
K. Reilly [1989] *The World And The West.* New York: Harper
A. J. Roque [1985] "Self-Organization: Kant's Concept of Teleology And Modern Chemistry," *Review Of Metaphysics,* 30: 107–135
——— [1988] "Non-linear Phenomena, Explanation and Action," *International Philosophical Quarterly,* 28, no. 3, pp. 247–255
R. Rosen [1985] *Anticipatory Systems.* New York: Pergamon
S. Salthe [1985] *Evolving Hierarchical Systems.* New York: Columbia
A. D. Schmookler [1984] *The Parable Of The Tribes.* Boston: Houghton Mifflin
J. Schull [1988] "Intelligence And Mind In Evolution," *World Futures,* vols. 23–4, pp. 263–274
C. E. Shannon and W. Weaver [1949] *The Mathematical Theory Of Communication.* Urbana: University of Illinois
——— [1951] *Bell Systems Technical Journal*
M. R. Stein et al., [1960] *Identity And Anxiety.* Glencoe: The Free Press
A. Toynbee [1961] *A Study Of History.* Oxford: Oxford University
V. Turner [1986] *The Anthropology Of Performance.* New York: PAJ
——— [1969] *The Ritual Process.* Chicago: Aldine
H. H. Turney-High (1971) *Primitive War.* Columbia, S.C.: Univeristy of South Carolina
R. Ulanowicz [1986] *Growth And Development: Ecosystems Phenomenology.* New York: Springer-Verlag
——— and B. M. Hannon [1987] "Life And The Production Of Entropy," *Proceedings of the Royal Society* B232, pp. 181–192

——— [1987] "On Quantifying Formal And Final Causality In Ecosystem Development," Presentation To Washington Area Evolution Club

A. P. Vayda [1968] "Primitive Warfare," *Encyclopedia Of The Social Sciences*, XVI

G. A. Wells [1987] *The Origin Of Language*. LaSalle: Open Court

B. L. Whorf [1956] *Language, Thought and Reality*. New York

P. Wohlmuth [1988] "Nested Realities And Human Consciousness: The Paradoxical Expression Of Evolutionary Process," *World Futures*, 25: 199–236

Q. Wright [1942] *A Study Of War*. Chicago: University of Chicago Press

CHAPTER 7

Economic and Social Evolution: The Transformational Dynamics Approach

PENTTI MALASKA

Editors' Introduction: In this study Pentti Malaska, a globally minded scientist and management consultant, opens still wider vistas of social change. The perspective is that of humanity itself, in the time-span of the last several thousand years. The "transformational dynamics" of Malaska is an application of the Prigoginian dissipative structure theory of Artigiani to the understanding of worldwide development. In the here espoused extended framework, the revolutions discussed by Artigiani become elements of an ongoing development process through which leading-edge human societies can transform their will from their tribal origins and their present—and foreseeable future—industrial condition to a novel vista of a post-industrial condition.

The historical panorama sketched by Malaska highlights the role of major bifurcations: those Artigiani would rightly consider "catastrophic" rather than "subtle." In this analysis, bifurcations are introduced into social processes because of the inability of the faculty of the dominant social and economic orders to facilitate new emerging needs. The interaction of mental and ethical elements of reality such as needs and values, with material elements such as the social and material order of society, is the holistic sphere that produces social change, and creates the evolutionary dynamic propelling humanity from the past toward the future.

The thesis intends to elucidate the nature of the choices that await contemporary people and societies as they head beyond the industrial stage. The distinction between societies of basic, tangible, and intangible needs, as that between extensive and intensive growth, are potentially valuable tools of social science analysis. A conceptual framework rather than the systematic application of a finished theoretical scheme, the ideas brought to light here would be fruitfully debated by historians, economists, development experts, and futurists alike.

INTRODUCTION

Volitional Change

The world is experiencing a period of change the prime agent of which is the development it has brought on itself, in particular the growth of industrial activity that has taken place since the last World War. The quantitatively replicative nature of the growth of the industrialized economies, increasing number of economic interactions in the world and the unforeseeable dissipation of nature which it caused has brought about major change. The world has become different: continued growth in the experienced extensive sense is more difficult. Many people in the industrialized world no longer fully believe that further development can be ensured by a policy of "gaining weight" by material growth. However, great disparities exist between nations as well as between people within some nations to an extent that there is a real and sustained lack of satisfaction of even basic needs.

It seems that during the 1970s, futurologists, whether they were ultraoptimistic as the late Herman Kahn or ultrapessimistic as Robert Heilbroner, found at least a common denominator for the future (Cole 1978).

Three characteristics of future visions, from Kahn to Heilbroner, are found to be common and similar to that of The *Limits to Growth*. First, that we are ending the zero growth state of material well-being. Secondly, that this state will be reached within the next 100–200 years. And thirdly, that the present generation is living at a turning point where the overexponential growth in the developed world will begin to slow towards zero growth.

This, however, does not reveal all the qualitative aspects of visions, especially those that are incommensurate with the one expressed by growth. The end of growth does not mean the end of development. It only means that in development some new qualities will take the dominant place of quantitative growth while others will stay constant or even diminish. Previous measures of the standard of living will loose ground and new measures, will be needed to demonstrate the continuation of development. However, development will always remain only as a possibility among many other non-development options of life. Development is not automatic; it is the result of conscious and wise choice.

The 1980's have shown us the disparity of the least developed countries in the world community, discouraging belief in automatic development (Lemma-Malaska 1989).

The expectation of, and demand for, something new has now entered the human spirit. This results in the need for finding new meanings of the concept of growth as part of the process of human evolution to enable us to see the possibilities of a "slim quality of life," and to set them as valuable goals for human activity. The purpose of human life cannot be defined by material terms however severe the lack of these basic resources, may be. In any case; the widening of the gap between the haves and the have-nots must be regarded as an implication of non-fullfillment of the purpose of human existence. A demonstration of increasing interest in this issue is the discussion and debate inspired by the Club of Rome during the past fifteen years. The second report to the Club of Rome *Mankind at the turning Point* (Mesarovic and Pestel, 1974) drew already our attention to the need of renewing growth. Since then the concept of organic growth along the long tradition of biological analogy in social thinking came to be widely accepted in discussion. In transformational dynamics we aim to go further, and insert conscious, human volition in the model of development. The idea of volitional growth includes in the organic growth prospects as well as necessary quantitative growth prospects. "The real issue is how can the rich be convinced that they should live more simply in order that the poor may simply live" as one of the Club of Rome members put it at the Club's Helsinki conference (Thapar 1984).

The role of technology to support overall development is vital, but it is not as unproblematic as commonly held. (I have discussed this subject in another paper and refer to it for further details (Malaska 1986).)

The concept of growth has many dimensions and thus we need not give up the use of the concept itself but only the unidimensional interpretation of "gaining weight," or its meaning as extensive growth. Instead, attention will be drawn in this paper to the meaning of "intensive growth" and "regenerative growth" as necessary elements of the transformational dynamics of development.

A plethora of new phenomena of change with world-wide significance have arisen since 1973 (Malaska 1983, 1984, 1985). We are experiencing today a change of vital parameters characterizing economic functions, social values and environmental conditions and a change in the mutual interactions and links between these issues within single nations as well as between all the nations of the world. The world as a system seems to be shifting away from its previous steady state towards a possible bifurcation, the long-term importance of which is comparable to that of the industrial revolution. In this process of turbulent change and structural instability some people and nations may gain and some may lose, and the causal relations between them will change. Some of the rationales previously prevailing will lose their ground and new ones will appear, with new entities, links and overlaps.

Changes in the environment can be met in two ways: either by learning to understand new conditions and changing one's own activities correspondingly, or by giving up the search for further development and the purpose of life.

This study deals with the first alternative: the continuation of development of societies. In this search the successive patterns of development are classified as societies of basic needs, tangible needs and intangible needs. The theory of complex nonlinear, self-organizing systems far from equilibrium serves as a general logical framework (Nicolis 1977, Prigogine 1980, Allen 1984, Laszlo 1985). It is called here transformational dynamics.

Development

Each new stage of the development has within it the seeds of further change. This is a basic idea of the transformational dynamics and it also underlies evolution and feedback (Fig. 1). Accordingly, development means self-organizing changing orders emerging as a result of nonlinear nonequilibrium processes triggered by local fluctuations, and not merely of perennial global equilibria.

The magnitude of human activity, compared with the scale of the natural ecosystem, has already reached a level where the interaction between man and nature cannot be ignored. (Malaska 1971, 1985). The onset of nonequilibrium can be triggered by comparatively small local fluctuations, either originating within the local subsystems or coming from the outside. Once established, the fluctuations become amplified and spread in the domain of the subsystems. Then they constitute a sizeable force capable of modifying macrobehavior. A mechanism for local nucleation and fluctuations is thus vital. In too large a system, small fluctuations which occur naturally are easily damped before they can affect average behavior, whereas fluctuations which are large enough occur too rarely to be considered (Prigogine 1972).

In this study I put forward the hypothesis that developmental nucleation can only materialize around some needs so far left unsatisfied. The mode of production (agricultural, industrial, etc) is merely a manifestation of changing material orders to fit with the desire to satisfy such needs.

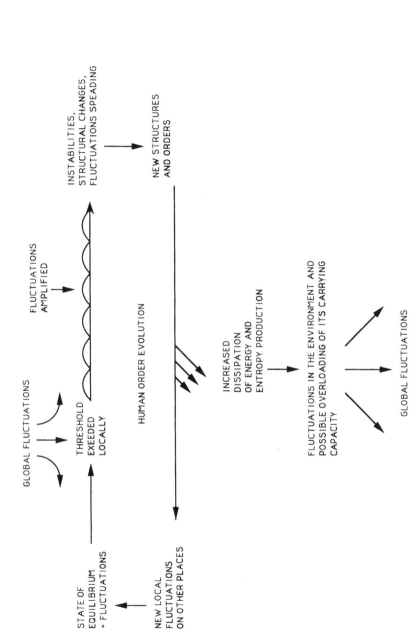

Figure 1. Cultural evolution feedback and related global problematique. Reproduced, with permission, from *Africa Beyond Famine: A Report to the Club of Rome*, Tycooly Publishing, London and New York, 1989.

This means that development is to be regarded as the process of inter-action between the mental (needs, values), social, and material order of the world. (Malaska 1977). Neither of these orders can determine the future alone; all appear as essential preconditions for harmonious interaction.

The basic difference between chemical, biological, ecological, and human systems is that in the latter the impulses causing primary fluctuations are initiated not only by chance but by man himself, and may be initiated consciously. These sources include learning, innovation and invention, entrepreneurship, citizen movement, and even rebellion; they also include the ceaseless devastation of nature and the environment and wars and threats. Technology is evidently one of the main functional, compositional and structural fluctuation within the world system, in a both positive and negative sense. Important aspects are, however, also changes in values and political turmoil. All of these effected from the 1970's onward a mingling of orders of different kinds, economic and ecological, national and regional, etc.. This has led to an increase of complexity in the world as a whole.

The future will not necessarily take shape as we shall outline here. There may be people who do not desire development of the type de-scribed but prefer a continuation of tangible material growth as before. However, I believe that the here offered outline is both desirable and feasible, particularly as a solution for which, by means of information technology and new scientific knowledge, the material basis can be en-sured. However, the widely used term "information society" is not ad-equate to describe these new possibilities since it only draws attention to material factors of development and ignores other, more vital ones. The changes in our very needs and values are the more crucial issue. The term "information society" may be used in a restricted sense to mean the intensive growth stage of industrial society (the society of tangible needs). For this reason, a better term for development is "the society of intangible needs."

Future Thinking

How we see the future is based largely but not solely on free and cre-ative thinking. To be realistic, future alternatives must be sufficiently identifiable as derivatives of the present and the past.

Future thinking by analogy, enriched by other means (such as utopia, dystopia, trend and scenarios), is adopted here (appendix A). It is also necessary to point out the limits of conventional "systems thinking." Systems thinking with changeable boundaries of systems is what is really needed (Checkland 1981). The renewal of boundaries is, if not the basic aim of development, at least its result.

MAJOR BIFURCATIONS IN THE PAST AND THE FUTURE

Bifurcations and World Problems

If we take the beginning of industrial society as the time when industrial production was getting off the ground, then it is about 100 years old. It is only a few decades since industrial production began to dominate and shape the lives of people in societies. The industrial world evolved in bifurcation from the agricultural world. However, some nations have never reached this point of bifurcation and perhaps never will. It was not until the mid-1960's that the (monetary) value of industrial production outgrew the production of other sectors in the world economy.

The term "post-industrial society" assumes a major bifurcation from industrial to a new kind of a society, differing from industrial society as much as ours differs from the previous agricultural one (Fig. 2). A large number of developing countries have only recently become industrialized and now they face this new bifurcation. And how can societies which are poor, being societies of basic needs, benefit or suffer from these world-wide discontinuities?

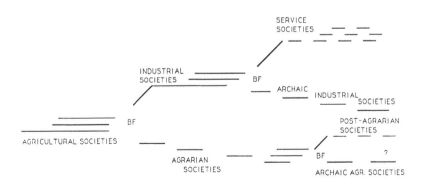

Figure 2. Diversification of societal development and major bifurcations (BF). Reproduced, with permission, from *Africa Beyond Famine: A Report to the Club of Rome*, Tycooly Publishing, London and New York, 1989.

What will happen to industrial societies that are unable to reach the bifurcation stage, and to those that are still agricultural? I will not try to answer these questions, although they are just as important as post-industrial development itself (Cole 1984). Today the industrialized world, and in particular a few of its most advanced industrial societies, dominate the other societies of the world. It is difficult to say whether this overall dominance will be maintained in the post-industrial world, or whether development will produce an essentially novel and more just "power order" different from the past and the present.

Human evolution is constrained, affected and coordinated by a few universal and global phenomena and tensions that operate like field phenomena and not merely like local causes. There are four main global fields—of which we are only in partial control—which are important features of development. They are: (1) East-West antagonism, (2) North-South imbalance, (3) Man-Nature imbalance, and (4) the force of technological development (Peccei 1981). Local fluctuations may be triggered, amplified or dampened by the presence and control of the field in a way that is difficult to foresee (Nicolis 1981).

Germs and Nucleation

Any major bifurcation requires as a necessary requisite a germ to start the nucleation process. The germ serves two purposes in development. First, it is needed to make a contribution to the dominating production mode, in particular to increasing its productivity and efficiency. This germinating activity does not emerge from the dominating production mode, but from outside of it; it is very different from it. By producing new means (hardware, software and orgware) for the dominating mode, this cross-catalytic effect transforms the dominating sector from a stage of extensive growth to one of intensive growth. During the period of intensive growth, wealth and welfare are accumulated and can be used to satisfy new needs. These new needs stimulate nucleation in the development process. The other function of the germinating activity is to auto-catalytically organize itself, to grow extensively and finally to take the role of the dominating production mode in society in satisfying new needs. This chain of development, and the shifts of the different types of growth, are illustrated in Fig. 3. According to their nuclear needs, societies are classified as societies of basic needs (SBN), tangible needs (STN), or intangible needs (SIN). Differences between the various stages

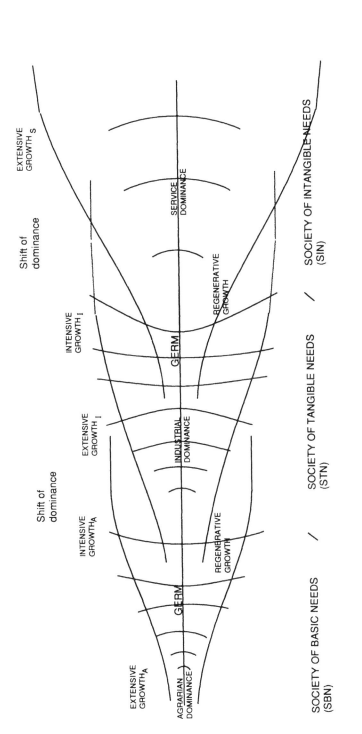

Figure 3. The transformational dynamics of societal change. Reproduced, with permission, from *Africa Beyond Famine: A Report to the Club of Rome*, Tycooly Publishing, London and New York, 1989.

Table 1

COMPLEX GROWTH

(0)	Autocatalytic germ emerges and regenerating growth may start
(1)	Extensive growth

Objective:	as much, as fast, to as many as possible the means of needs satisfaction
Policy:	extensive exploitation of resources
Effectiveness:	gross production
Measure of standard of living	resources used per capita

(2)	Intensive Growth

Objective:	more from less, better than before, entropy efficiency
Policy:	increase in resource efficiency and in quality of products and services
Effectiveness:	productivity in use of resources
Measure of standards of living	gross production per capita

(3)	Transmutations of the previously dominant sector

of complex growth—extensive, intensive, regenerative—are described below.

EMERGED SOCIETIES

Development as a Process

In Fig. 4 the process of development of emerged societies is illustrated in a more detailed fashion. The arrows marked (1) indicate the formation by renewed growth of a new dominating production sector from the preceding germinating activity. The germ is created in the first place to serve the prevalent production mode and its increased productivity.

In a society of basic needs the new germ fuels the further development of agriculture in the form of embryonic industry and related functions. These include the manufacture of tools for agriculture, clothes and other goods from agricultural raw materials together with the use of information for cattle breeding and plant improvement and the subsequent development of working methods. By contrast, information technology, and new scientific knowledge (in a broad sense) instead of manufacturing, constitute the germ in societies of tangible needs.

Another point is illustrated by arrows marked (2) which form cycles. The cycles describe an autocatalytic interaction between the dominant-production sector and the germinating functions. Thus the dominant sectors are shifting from a state of extensive to intensive growth, away from equilibrium. Intensive growth creates new wealth and a new distribution of wealth. Fluctuations emerge as a result.

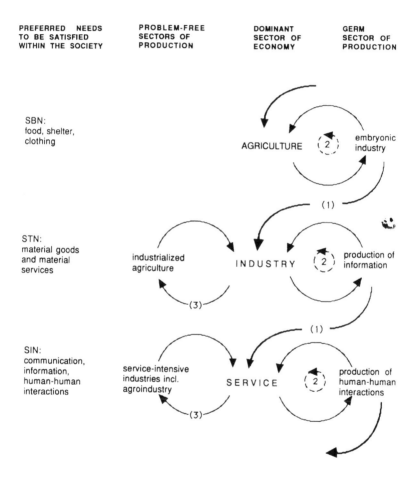

Figure 4. The processes of societal transition: Arrows marked by (1) indicate the formation of a new dominating autocatalytic production sector from the preceding germinating activity; shift of dominance. Arrows marked by (2) describe a crosscatalytic interaction between the dominant production sector and the germinating activity. Arrows marked by (3) indicate the change in dominant position between the prevailing and emerging production sectors.

Due to fluctuations and new wealth produced by intensive growth, new needs are manifested and are beginning to be satisfied. Basic needs no longer remain unsatisfied. The society of tangible needs is born and industry has the potential to serve it. In the future, tangible needs will become satisfied and lose their dominant position to new, intangible needs. The extensive growth of industry (arrow (1)) brings about a renewal of the agricultural society of basic needs and transforms it into an industrial society of tangible needs.

If we assume that this dynamic also holds in the future, we may expect a new type of society, a society of intangible needs to emerge after today's industrial societies. The germ of this transmutation is information, new scienfic knowledge, and better human relations and associated needs.

The arrows marked (3) describe the change of the dominant position between the prevailing and emerging production sectors. Agriculture becomes a special branch of industry which in turn takes the dominant position at the first bifurcation. At the second, industry follows agriculture and also becomes a non-problematic branch of production in the post-industrial society of intangible needs. The new dominant production mode for intangible needs is the service economy.

THE SOCIETY OF BASIC NEEDS

For purposes of this study, the society where traditional farming, forestry and husbandry, that is agriculture in the broad sense, is the dominant force of development is described as the society of basic needs. I believe that the development phases of society are not determined by methods of production as such, but by new needs, the satisfaction of which is considered to be the primary goal of society. The organization of production and the realization of consumption to satisfy these needs are the focal points of any existing economic and social policy.

In traditional agricultural societies, this nucleation of development occurs around the basic needs for food, clothes and shelter. The satisfaction of the primary needs for food, drink and shelter is regarded as the objective of society. Organizing and implementing this objective entails traditional farming, cattle-raising, and forestry.

The problem of fulfilling these basic social needs must be solved in the most efficient way possible. In this process evolve suitable production methods, infrastructure, concepts of work and livelihood, family composition, methods of upbringing and care, formulas for the possession and exercise of power, and even particular social values. The entire society, including its citizens, assumes the most appropriate form to ensure the satisfaction of needs. In addition, a power hierarchy is created among the various suborders, the apex of which is the dominant mode of production. At the apex of the society of basic needs is agriculture, with farmers and landowners. The agrarian way of life, work and rationale are the criteria for almost all activities.

Livelihood and the concept of work in agrarian societies is society-specific, as is the idea of the family. The family is a unit of both production and consumption, of maintenance and education, and therefore it is bigger than in modern societies. Male and female roles are also different. This is advantageous and effective from the point of view of meeting the primary goals of society, that is, basic human needs. Contemporary ideas on these matters are the "special products" of industrial society, created by its particular requirements for efficiency in satisfying tangible needs.

Extensive Growth

Initially the most challenging problem of a society of basic needs is to increase production by expanding the resource base, the land area and the number of cattle. For a long time, solutions were based on a policy of extensive growth: ever more land and cattle, more forest for productive use. The phase of extensive growth lasted thousands of years. For the majority of mankind, it is fundamental for the improvement of the quality of life even today. However, for the industrialized world it has not sufficed for the past decades. Cultural evolution has bifurcated with the emergence of industrial production.

At the stage of extensive growth of a society of basic needs, human well-being is measured in acres of tilled land per capita.

Intensive Growth

Gradually materials (fertilizers), tools (implements made in machine shops, machines driven by animals), and other manufactured products (for dairies and mills) produced outside the dominant sector of agriculture were introduced and utilized by farmers. At a time when agriculture was still in a preeminent position in society this outside contribution by manufacturing enabled agricultural production to become more effective and productive in the use of resources and the utilization of its products for consumption. While the germ was not agriculture, it affected great leaps in agricultural productivity.

However, the real work of society was still performed in the fields and in forests and not in manufacturing plants. Outside contributions were justified only by improvements in efficiency and the attainment of versatility for the producers. The society of basic needs remained an agrarian society in its dominant production mode and its values. However, its form of production changed from a state of extensive to intensive growth. It was through a continued increase in the use of inputs from outside that this transfer occurred. Simultaneously, intensive growth created a material base for new development. The whole of society moved toward a point of bifurcation; fluctuations, and a nucleus of new needs different from basic needs were formed.

The policy of intensive growth aims at getting more from less, that is, more and better products to satisfy basic needs, without increasing the use of resources.

The intensive growth period of a society of basic needs results from the mechanization, chemicalization, breeding and treatment of plants and animals, and a division of labor. During this period it is no longer necessary to till more land or add to the number of cattle at the same rate as before. Every acre and every animal yields more than before. Work and the distribution of the results of work also change. Workers may be classified as producers, persons employed but not really needed, unemployed and free persons, and renewers. The last-mentioned are at first incorporated with the others but later are clearly distinguished as workers in new areas producing the outside contribution for the intensive growth phase.

At first intensive growth accumulates new wealth mainly for the producers in the dominant sector and not for the rest of the society and for all consumers. Wealth is the means by which landowners, owners of capital, and workers in remunerated jobs achieve higher incomes. This may happen at the cost of the rest of the population, for example, when production is intensified by the replacement of labour with machines. The share of work is decreased by increasing the surplus amount of capital. This kind of substitution is always limited in its ability to foster good development and benefit society as a whole. Support for it is not part of a correct policy for intensive growth.

Intensive growth means the application of the idea of "more by less" in such a way that the inputs of all resources are simultaneously decreased. The measures of well-being in the period of extensive growth (such as acre per capita), must be replaced by another yardstick, for instance, by quantity of food per person produced.

Regenerative growth

When agriculture reaches the regenerative stage, excess material and social wealth accumulate in correspondance with savings in inputs and costs. In addition to making production still more efficient, this additional wealth is used to an increasing extent to raise the basic consumption level of the poor and to satisfy new needs. The latter no longer have any connection with agrarian production and they cannot be fulfilled by farming and animal husbandry. The term "regenerative growth" is used for these new needs as they emerge and begin to be satisfied by the products of formerly ancillary activities. The new needs satisfiable by manufacturing are called tangible needs.

The policy of regenerative growth is to exploit new needs by enlarging production and organizing consumption. The means for satisfying tangible needs are different from those of basic needs. Tangible needs require commodities, goods, and material services.

The society of basic needs changes into a society of tangible needs through regenerative growth, and becomes a society of industrial production and mass consumption. The satisfaction of basic needs does not lose any of its importance; these needs always keep a key position. However, the problems connected with the organization of the production and consumption required for their fulfillment are eliminated. Nowadays the industrialized countries have no difficulties in producing and offering to people any amount of varied foods according to their needs. Overproduction, and a labor force "superfluously employed in agroindustry" are the main problems.

THE SOCIETY OF TANGIBLE NEEDS

Intensive growth within the society of basic needs is based on contributions from outside, which have made agriculture more efficient. Here lies the cardinal difference between extensive and intensive growth. Extensive growth calls for ever more agriculture. But intensive growth within agriculture means that more and more of the total production value comes from sectors outside agriculture. The outside sectors contribute to agriculture and also provide the means by which additional accumulated wealth can be channelled into meeting new kinds of needs. The contributing sector embraces a seed from which regenerative growth outside the dominating sector begins, and which, in time, develops into the dominating sector of the next evolutionary stage.

Change of Dominance

The dominant form of production in the society of basic needs, agriculture in broad terms, ceases to be a major development problem. At the same time, the urgency to organize all of society anew, according to new criteria of efficiency, is recognized. People experience new degrees of freedom; new needs and problems arise.

The society coming after the society of basic needs is the society of tangible needs. Its prevalent form of production is industry rather than agriculture. Industry is born from the additional wealth brought about by the increase of efficiency in agriculture, following the phase when an embryonic industry only contributed to agriculture. Since then the industrial mode with its criteria for efficiency and functional requirements has taken a determining position in society. It dictates and defines in turn infrastructure, the concept of work, living conditions, family composition, male-female roles, the organization of education, health care, and even agriculture. It wields power and sets values. Today's values are as industrial as today's products.

For the maximum satisfaction of tangible needs, a maximum consumption of commodities is required. The immediate family and the one-parent family ensure the greatest possible number of consumer units, and maximum consumption by each. These units are the very insignia of the society of tangible needs, much as the enlarged family was the symbol of the society of basic needs.

In the society of tangible needs, goods are produced most efficiently by industry, and not by craftwork as in agricultural society. Industry and industrial progress facilitate the more immediate satisfaction of the tangible needs of ever more people. We have experienced a period of strong extensive growth in industrial societies which did not spare resources. We are more or less resigned to its reality-concept and to the values it created. The expansion of industry is all important; we are not yet prepared for any other kind of development.

At the extensive growth stage in the society of tangible needs GNP per capita can be used as an indicator of living standards. As long as productivity does not increase too quickly, GNP per capita is a good indicator of what society values most highly—the flow of goods and the extensive use of resources.

The Intensive Growth of Industrial Production and Renewal

Is there any catalyst in sight by means of which industrial production could move fairly quickly to the stage of intensive growth: to produce more from less to save capital, labour, raw materials, energy, space, environment and at the same time also improve quality and service? These new kinds of solutions are already available, although traditional thinkers in industry still seek opportunities to economize in one at the cost of others. Yet only non-substitutive automation can foster intensive growth.

Industry is on the verge of a stage of intensive growth. The intensive growth stage is reached when a rise in overall productivity is almost ineffective in increasing the production of goods, and only increases organizational costs and inflation. This happened in the 1970's (Voge 1983). As a consequence, society is approaching change from a period of tangible needs through regenerative growth to a radical new development, a stage where new needs prevail together with the new production methods they inspire.

At first this renewal process appears to be the work of a non-industrial contribution to industry. The objective is to accelerate total productivity in industry and thereby create additional wealth. The overall value of industrial output—the quantity, quality and availability of goods— increases and it becomes easier and easier to manufacture and obtain goods for consumption. The role and value of industry and its ability to renew will be considerably enhanced before it will yield its dominant position to a new production mode for the satisfaction of intangible needs.

Continued overproduction entails an intense search for new outlets and, simultaneously, radical changes in the numbers of the employed, the unemployed, and the renewers. GNP per capita no longer gives a real picture of living standards; a new measure (or a continuous adjustment of the content and weights of the measure), is needed.

The additional accumulated wealth can be used both for further productivity improvement, and for the satisfaction of the tangible needs of the poor. It can also be used to regenerate growth, that is, to open up possibilities of fulfilling more intangible needs.

THE RENEWAL POLICY OF ORGANIC GROWTH

Non-Substitutional Automation

The policy for the future is a policy of adaptation by means of intensive growth and the renewal of needs. The time span required for setting goals, allocating resources and getting results is longer than any parliamentary cycle. Hence the future will require special treatment by policy makers.

The catalyzer for industry in a period of intensive growth is information, scientific knowledge, and the development of human relations. The nature of information technology is so general that it can be applied everywhere and at all stages of industrial activity—in production, distribution, and consumption. Nowadays information technology has an established uniform conceptual basis (digitalization), and a uniform material basis (electronics) in all areas and functions of information services. This applies to information for data, picture, word, sound, colour, form, and to information transfer, handling and presentation (Kobayachi 1982). However, it is vital to see the concept of information more broadly than it is usually perceived, to tackle it with its full semiotic span (syntactical, semantical and pragmatical), and also to take into account new scientific knowledge from other fields. Information technology alone is not sufficient to bring about any revolution. New scientific knowledge can lead to savings and improvements. The materialization and implementation of such knowledge contribute to the accumulation of wealth. However, the process requires not only a new distribution of work and income (Leontief 1983), but a new understanding of work and earnings (Giarini 1984, 1989).

The first task in renewal policy is to search for ways to achieve non-substitutional automation and the novel implementation of information technology and other new scientific knowledge. It is to rationalize production, develop new products, and replace the concept "cost added at factory" by "service value at the customer's " as the decision base. It means building up an infrastructure based on an economy of knowledge—the supply of know-how and information services (Kyläheiko 1983). It also requires the stopping of subventions for branches of industry not equipped for intensive growth. The problem can no longer be solved by supporting overproduction. It is important to internationalize and establish new market positions. The trend is towards greater interaction, national and international, and sustainable cooperation.

The Society of Intangible Needs

Information and information technology are just as important for the satisfaction of intangible needs as power engines were for the satisfaction of tangible needs. Additional wealth can be channelled into the fulfillment of new needs by means of a new kind of production potential created by the contributing activity.

Information technology is thus a part of the material basis on which the opportunities to meet new needs are founded. It is a vital part of the infrastructure, but it is not the only vital element of society. Therefore, there is not enough justification for calling the next stage of society the information society, just as it would not be right to call the present stage a society of cars or jet engines. The term "information society" belongs to the intensive growth phase of industrial society. The dominant production sector of the society of intangible needs may also be the service sector, and the society of intangible needs may also be called "service society."

Table 2

NON-SUBSTITUTIONAL AUTOMATION

The purpose of automation is to make people free from the work of producing material goods and free for activity of a more human kind.

Non-substitutional automation generates real growth of wealth and not only the distribution of wealth as substitutional automation does. Non-substitutional automation is a means of intensive growth and at the same time a part of the germ function.

Criteria for non-substitutional automation:
    It saves
        materials
        energy
        work
        capital
        costs
        environment
    It improves
        quality
        service
and all of these simultaneously, non-substitutionally.

Goals

The prime objective in any particular stage of social development is to fulfill some determined needs. I consider that both needs-oriented and technology-oriented assessments of development point in the same direction, namely to needs that can be only satisfied together with other people in a variety of human relationships—at work, study, in families, and in micro societies. Thus the terms "society of intangible needs" or "society of interactive needs" are justified.

Industry offers the most efficient method of producing goods, and it will do so also in the future. However, there will be a change in the way it functions: it will be built around domination by human interaction instead of by machine systems. However, human interaction and communication cannot be produced by industry. There is a great "gap of intangible needs" in today's advanced societies of tangible needs. This gap cannot be filled by information technology. We must learn to develop human interactions as objects to be produced and consumed.

Creative work will change within industry and soon move away from it. Dynamic changes involving the employed, the unemployed, and the renewers take place within the dominant sector. At present, people freed from work in the production of goods include those who are no longer enthusiastic about, or capable of, meeting commodity production requirements (the retired, the sick), those who have as yet no proper experience of work and have no personal concept of it (the young), and those whom enterprises no longer find profitable to employ. In the near future the main purpose of work will be to offer people more and more ways for personal and social development.

Renewal policy should aim to release from commodity production a sufficient number of people who are disposed to engage themselves in regenerative growth. One way of encouraging this would be to alter the system of personel revenue, so that all citizens would receive a part of their income independent of work in goods production, just as the unemployed and retired do now. Another way is to give sufficient encouragement and incentives to people to apply their own entrepreneurial skills in the area of new needs.

Moreover, material and social conditions should be created to cater for new types of family structures in the society of interactive needs. In such a society there will be more family types than exist now, including new, larger families not held together by kinship. Family functions may be extremely varied ranging from maintenance and care to the provision and development of close human relations. All this could trigger creative activity in people, which in the last analysis is the only real resource for development. The new science of complexity will be required to understand these phenomena (CoR 1984, Prigogine 1984).

Value Capacity

The debate about values has deep and multifarious connections with the ongoing phenomenon of change. To what extent can we consider values as rigid—must they always remain unaltered? Or is it possible to change values? To what extent are they alterable and to what degree a matter of choice? Do we possess this type of value capacity?

On the other hand, values are closely and thoroughly linked to needs. When we are bound by material things, we reflect them in our values.

Table 3

THE PROGRESSIVE CLASS DIVISIONS OF SOCIETY

Producers
Persons engaged in regular paid work whose presence or activity at a given place is
necessary to achieve or ensure the achievement of a productive or administrative result
against payment.

Other employed persons
Persons engaged in paid work whose presence or activity at a given place could be
reduced, transferred to other areas or entirely eliminated by the rationalization or re-
arrangement of work.

Unemployed
Persons who are engaged in seeking productive work or other employment or who are
being trained, re-trained or rehabilitated for productive work or other employment. Free
persons are those not seeking employment, nor undergoing training or rehabilitation in
the above sense.

Renewers
Persons engaged in productive or other employment, or who are free, and who receive
their income from contributions to the intensive growth of the dominating sector or by
their activity or lifestyle; such persons create the necessary conditions for new needs
and the regenerative growth of society.

Our present values are different from the values of agricultural society;
they are products of industrial society. Our values are in part imposed by
industrial society to safeguard its own functioning and continuity. This
can be clearly seen, for instance, in our way of stressing the importance
of having an abundance of material goods and commodities for our personal use.

If we accept that a capacity exists to alter our values we can move
more easily to a new society where things other than material values and
the fulfillment of tangible needs are the major objectives. Such objectives can be the intangible needs of human relations, human interactions,
better communication, etc. Instead of concentrating on the satisfaction
of additional tangible needs, we could focus on the intangible needs of
human relations. At present the insufficiency gap is widest in this very
area (Milbrath 1989).

APPENDIX
WAYS OF VIEWING THE FUTURE

Utopian Thinking

This method of thinking is to subject the present social situation to an
analysis of the possibility of satisfying certain new needs. Because of the
novelty of the target needs, this way of thinking is called utopian. By

utopian thinking it is possible to find need 'gaps,' in other words desirable but so far unsatisfied needs.

Examples of utopian thinking are Francis Bacon's *New Atlantis,* Thomas More's *Utopia,* Botkin-Elmadrja-Malitza's *No Limits to Learning,* and Toffler's *The Third Wave.* Among such needs of vital importance are new kinds of inter-personal, emotionally positive interactions and needs that can be satisfied only in human communication. Our present relationships and our present moder of communication provide but an insufficient indication of what may be involved. These needs become more evident only when we satisfy them.

Dystopian Thinking

The present situation can be also viewed as an enforced state compared with a better one. The result of such a comparison is a catalogue of the shortcomings of present society. The abolition of these shortcomings (i.e. the elimination of the existing negative situation) reflects the existence of special needs. Attempts to satisfy the needs can be considered efforts to create a better future.

Examples of today's shortcomings are the environmental problems arising from the technological and material progress of the industrial countries and the poverty of developing countries, the wasteful exploitation of natural resources and neglect in conserving such resources, and the plunder of certain sections of world society. The needs reflected in these phenomena cannot be satisfied by continuing and intensifying past trends towards material growth and technological development. Instead, this trend must be broken and new directions for technological development must be found. From this, new views for directing growth and technology will arise. Examples of this kind of thinking are the Club of Rome report *The Limits to Growth* and Orwell's *1984.*

Analogy Thinking

To characterise the future by the analogical approach it is necessary to take a longer period of time, a span for which there is already a 'description' or 'explanation.' Agricultural society and industrial society and the break between the two were the above examples. By placing industrial society in the situation of agricultural society at the time of the break between them, it becomes possible to use the special features of the break to explain the emergence of new needs, of regenerative growth, and the evolution of a new dominant social force.

The process of analogical thinking adapted to the description of the future may be illustrated by the following diagram:

Problem:

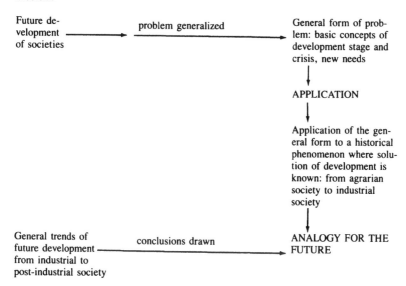

Future development of societies —— problem generalized ——→ General form of problem: basic concepts of development stage and crisis, new needs

↓

APPLICATION

↓

Application of the general form to a historical phenomenon where solution of development is known: from agrarian society to industrial society

↓

General trends of future development from industrial to post-industrial society —— conclusions drawn ——→ ANALOGY FOR THE FUTURE

Perhaps the best-known exponent of analogy thinking for describing different futures is Herman Kahn in *The Next 200 Years.*

Trend Thinking

The trend approach was, until relatively recently, the best known and most frequently used method of characterizing and predicting the future. All statistical and mathematical methods regardless of their degree of complexity belong to this group. The trend mode of thinking is based on a known and invariable pattern. It is one of the most used tools in the futurologist's arsenal, even though an uncritical use of this method brings with it the unconscious assumption that the world is mechanical and regular.

There is no limit in trend thinking as to what feature of the world is selected as its most important characteristic if it is (or can be assumed to be) sufficiently invariable and regular. By trend is meant not only something which can be expressed in quantitative terms; it encompasses also other phenomena which may be regarded as unchanged, or as changing in the same way as in the past. An interesting example of trend thinking is Cesar Marchetti's studies of technological development at IIASA.

Railway Thinking

Linear railway thinking has been one of the most used methods in practical politics for describing and justifying the future. It operates on the simple belief that the (ordinarily desired and wished for) course of events in one country will be repeated in other countries in due course. Development is likened to a railway track, along which the trains of nations move one behind the other at intervals, each passing in turn the same stations and the same scenery. The future for those behind resembles the past of those in front. In some cases such thinking can scarcely be regarded as anything more than an indication of the lack of creative thinking. In others cases, however, it may be the result of material interrelations and cause-and-effect chains between different countries.

Scenario Thinking

Scenario thinking was first used in futurological studies in the 1950's but it was not until the 1970's that it became the most important tool for creating maps of the future. It is basically an intuitive approach which makes use of all the approaches already described in trying to create alternative views, termed scenarios, of the future. In constructing and selecting scenarios one can draw on such techniques as morphological analysis, the delphi technique, the cross-impact technique, the relevance-tree technique, and mathematical models.

In drawing up scenarios, attempts are made to construct a logical series of events which demonstrates how the future situation develops, step by step, out of the present. An active operator can examine each stage, or he can make his own adjustments at certain intervals and add his own ideas to the scenario. In the latter case, the scenario becomes his own subjective view of the future.

A scenario may be considered a set of multiple predictions; it is not a prophecy. A scenario may be regarded as planning information; sometimes its purpose is to cast light and reveal the quality and characteristics of the present by projecting it farther away from us and 'enlarging' it. Often scenarios illustrate the nature of the possible and how it can be achieved. In this way they prepare us for various alternatives.

REFERENCES

Allen, P. M. (1984), Towards a New Science of Complex Systems. The Club of Rome 1984 Conference, Helsinki July 1984.
Checkland, Peter (1981), Systems Thinking, Systems Practice. John Wiley, NY 1981.

Cole, S.—Miles, I. (1984), Development, Distribution and the Future. *Futures*, Oct. 1984, 471–493.

CoR (1984), Managing Global Issues. Reasons for Encouragement. Proceedings of The Club of Rome 1984 Conference Helsinki July 1984. Helsinki 1985.

Giarini, Orio (1984), Complexity in Economics in CoR (1984) above pp. 161–167.

Giarini, O.—Stahel P. R. (1989), The Limits to Certainty. Kluwer Academic Publisher, London, 1989.

Kobayachi Koji (1982), Man and "C & C": concept and perspectives. IIC Conference Sept. 1982 Helsinki.

Kyläheiko, Kalevi (1983), The Problem of Technological Unemployment. Lappeenranta Technical University. Lappeenranta.

Laszlo, Erwin (1987), Evolution: The Grand Synthesis Boston and London: New Science Library, Shamghala.

Leontief, Wassily (1982), The Distribution of work and Income. *Scientific American* Sept. 1982, pp. 152–164.

Lemma, Aklilu, Malaska, Pentti (1989), Africa Beyond Famine. A report to the Club of Rome, Tycooly Publ. London 1989.

Malaska, Pentti (1971), Prospects of Future of Technical Man. Insinoorilehdet Oy Helsinki 1971.

Malaska, Pentti (1977), New International Order and Technology, 4th Congress of Chemical Engineering 1977. Turku School of Economics A–3:1977.

Malaska, Pentti (1983), Coping with Unpredictability and Uncertainty of the Future. Turku School of Economics Publication A1:1983.

Malaska, Pentti (1987) Environmental Problems of Modern Societies. International Journal of Technology Management (IJTM), Vol. 2, No.2, pp. 263–268, 1987.

Malaska, P—Malmivirta, M.—Meristö, T.—Hansen, S-O (1984), Who uses scenarios— and why? *Long Range Planning*, Vol. 17, No. 5, pp. 45–49, 1984.

Malaska, Pentti (1985), Multiple Scenario Approach and Strategic Behavior in European Companies *Strategic Management Journal*, Vol. 6, No. 4, pp. 339–356, 1985.

Mesarovic, Michajlo—Pestel Edward (1974), Mankind at the Turning Point. NY 1974.

Milbrath, Lester (1989), Envisioning a Sustainable Society: Learning Our Way Out. Albany SUNY Press 1989.

Nicolis, G.—Prigogine, I. (1977), Self Organization in Nonequilibrium Systems. John Wiley 1977.

Nicolis, G.—Prigogine, I. (1981), Symmetry Breaking and Pattern Selection in Far-From-Equilibrium Systems Proc. Natl. Acad. USA Vol 78 No2, 659–663, Febr 1981.

Peccei, Aurelio (1981), One Hundred Pages for the Future. Pergamon Press 1981.

Prigogine, I.—Nicolis, G.—Babloyante, A. (1972), Thermodynamics of evolution. Part I. *Physics Today*, Nov 1972, 23–28. Part II. Ibid, Dec 1972, 38–44.

Prigogine, Ilya (1980), From Being to Becoming. Time and complexity in the physical science. Freeman, San Francisco 1980.

Prigogine, Ilya—Stengers, Isabelle (1984), Order out of chaos. Man's New Dialogue with Nature, Bantam Books, 1984.

Thapar, Romesh, A Statement to the conference of the Club of Rome 1984 in Helsinki in CoR (1984) above.

UNU (1985), The Science and Praxis of Complexity. Contributions to the Symposium held at Montpellier, France, 9–11 May 1984. UNU 1985.

Voge, Jean (1983), The Political Economy of Complexity. From the Information Economy to the 'Complexity' Economy. Information Economics and Policy North Holland, 1983, 97–114.

CHAPTER 8

The Evolutionary Paradigm and Neoclassical Economics

MIKA PANTZAR

Editor's Introduction: Kenneth Boulding once quipped about contemporary economics that "nothing fails like success." The failure of neoclassical economics to take into account evolutionary phenomena in social and cultural development may be due to the relative self-centeredness that is a frequent corollary of success. But, if in today's rapidly changing societal environment neoclassical economics' success is not to fade irremediably, cross-fertilization with the evolutionary current in scientific thought may prove to be necessary. For general evolution theory itself, the development of some variety of "evolutionary economics" would mean extending its paradigm to a new and vitally important field of social science. The greatest benefit may, however, accrue to contemporary peoples and societies. For the cross-fertilization of these modes of thought could provide practical economic guidance under conditions of rapid and fundamental change, including instances of basic socioeconomic transformation and economic contraction. With such phenomena neoclassical economics is poorly equipped to deal.

Pantzar, a young Finnish economist, assumes the major task of examining the possibilities of creating bridges between the evolutionary and the neoclassical modes of thought. He follows up Malaska's global vision of the economic and social development of humanity with regard to the technical issues of economic interest. The task is considerable and its difficulties are not to be underestimated: to the best of this writer's knowledge, aside from occasional explorations by systems-minded economists and economically minded systems theorists, an economic theory of man and society in evolution has not been seriously attempted.

Thus, Pantzar breaks new ground. Evidently, here he cannot do so in great depth. But even these preliminary explorations uncover interesting findings: they include basic conceptual contrasts and differences between the evolutionary and the noeclassical paradigms, and pinpoint the failure of the latter to account for evolutionary changes in its fundamental parameters. In this regard Pantzar's description of *Homo oeconomicus* in evolutionary terms is a noteworthy, though as yet sketchy, attempt at economic theory innovation. An extended version of this paper has appeared in Pantzar, 1986.

Clearly, Pantzar's exploratory notes deserve to be followed up with systematic research, both by open-minded economists, and by the more fearless and economically-minded of the evolutionary theorists.

157

INTRODUCTION

This paper attempts to assess the feasibility of linking the "evolutionary paradigm" to the description of an economic agent as assumed by neo-classical economics. It contrasts the standard economic approach with a view which emphasizes the discontinuous nature of development. Concretely discountinuous development will be discussed by referring to economic contraction. Most likely economic contraction leads to profound social transformations.

This study is not a criticism of economics. Rather, its purpose is two-fold. First, to critically examine the descriptive limits of the economic approach. Welfare economic questions are not touched upon in spite of the fact that is it obviously welfare economic and normative implications that the evolutionary paradigm would change the most. The limits of economics are evaluated by comparing a highly simplified (stylized) illustration of consumer economics with the manifest features of evolutionary phenomena. One particularly difficult phenomenon—economic contraction—is investigated and one of the foundations of modern microeconomics—the assumption that preferences are given or exogenous—is contested.

Second, the paper describes one potential starting point—the evolutionary paradigm—which could provide a fruitful basis for further analytical discussion in economics. Possible hypotheses arising from the methodological and metaphysical principles of the paradigm are stressed. The paper could be criticized as merely a repetition of the long-standing doctrinal dispute between the historical and the analytical schools ("methodenstreit" at the end of 19th century). To some extent it argues against the mainstream in favour of the historical approach. However, the author does not regard it as a repetition because of the obvious new possibilities that the evolutionary paradigm offers for studying society in a non-mechanistic way.[1]

The paper seeks, especially in the application of economic contraction, to develop fruitful hypotheses which might later be evaluated empirically. It should be noted that we are working in a highly specific area of research. Often the results are theory-specific because we cannot directly perceive complex phenomena without some intervening (theoretical) reasoning. When studying complexity, there are no pure "empirical facts." When evaluating the descriptive power of economics in complex phenomena an obvious difficulty is to find a reference point against which the results of the approach might be contrasted. Thus, to compare economics with the evolutionary paradigm is not necessarily to compare it with reality.

THE SCIENCE OF COMPLEXITY

"Basic features in these paradigm changes imply both a reaction against, and, more importantly, a going beyond fundamental canons of the classical Western scientific tradition, thus beyond what now appear as mechanistic, linear, closed, deterministic, reductionist, universalist paradigms of this tradition. There is a basic shift from simplicity to complexity, from structure to process. Instability, openness, fluctuation, disorder, uncertainty, improbability are introduced into scientific models of reality. Thus, the new paradigms concern non-equilibrium physics, dynamic open-systems, dissipative structures, the creation of order out of noise and disorder." (Ploman, 1984).

The philosopher Alfred North Whitehead emphasized that the task of philosophy is to reconcile permanence and change, to conceive of things as processes, to demonstrate that becoming forms entities that are born and die. It is these metaphysical ideas which form one basis of the evolutionary view (Prigogine and Stengers, 1984; 95).

Metaphysics (or ontology) has been convincingly defended by some contemporary philosophers (Bunge, 1977; I, 23–24). The situation is different among social scientists, especially among those who emphasize "social technics" and pragmatic values of scientific activities. "Scientific ontology" as Bunge (1977, I, 24) states "is a collection of general or cross-disciplinary framework and theories, with a factual reference, mathematical in form and compatible with—as well as relevant to—the science of today." Whether or not the evolutionary paradigm is *scientific* ontology is an open question. It is, however, useful for an economist interested in disorder to attempt to draw a parallel between this paradigm and economics. (For the relevance of metaphysical questions to scientific progress see Bunge, 1977, I, 23–24.)

One important but rarely presented point should be made. It has both analytical and political dimensions: If economics and the evolutionary paradigm are to be linked, the ideas of the former should in the first instance be evaluated not according to its descriptive or predictive power but according to its *congruence* with both our contemporary theories about the economy and our conventional wisdom (i.e. commonsense thinking of decision-makers), which is often implicitly based on the former. To understand and govern systems by linking different views is possible only if the metaphysics of the views do not contradict.

One example of the problems caused by such incongruence in global policy discussion occurred in the global modelling of the 1970s. As the influence of economists in the field of global modelling grew, concern diminished with global issues as conceived by the Club of Rome (Meadows et al., 1982, 100). By applying a new framework with more narrowly defined boundaries, the original concern for long-term global development almost disappeared. Neither was the linking of these two views necessarily advantageous for economic theory. It might be that large (i.e. data-rich) models with a simple structure are not appropriate for understanding a complex global economy.

Is the evolutionary paradigm also applicable to the study of social and economic dynamics? It is true that new perceptions concerning the dynamics between stability and instability—between inertia and change—are emerging in several fields as Ploman (1984) argues. Ploman, though, goes too far in connecting, for example, the time-geography of Häger-strand and the linguistic theory of Chomsky. Still, one could add to the list offered by Ploman a number of evolutionary views (for example, Boulding, 1978; Daly, 1980; Dosi et al., 1988; Guha, 1982; Hodgshon, 1988; or Malaska, 1985; Nelson, Winter 1982), applications of the theory automata (Roeding, 1977) and the theory of finite complex systems (see Albin and Gottinger, 1983 and Gottinger, 1983), all of which have interesting implications for economic research. With the aid of these new approaches it is possible to make economics simultaneously more abstract and more institutional than many economic theories to date (cf. Roeding, 1977). Let us remember, however, that evolutionary theorists have existed before, even in economics. Joseph Schumpeter (cf. 1947) and Thorstein Veblen (cf. 1904) are among the most prominent. James Duesenberry (1949) also stressed the irreversible nature of development, particularly in the case of consumption.

There are both internal and external tensions in the sciences arguing for the inclusion of evolution in the analysis. Here we concentrate only on internal development; the discussion of increasing global complexity and turbulence is left to the next chapter.

Mechanistic models in economics refer to the representation of the dynamics of physical systems. Such representation is deterministic and reversible and requires that physical systems do not change their identities over time. They cannot become anything radically new. Prigogine's approach, on the other hand, emphasizes physical systems that exhibit radical novelty. This suggests the need for a new foundation of physics, one whose relationship to identity, time and dynamics differs from that of classical mechanics. We argued above that in neoclassical economics inner and outer environments are given and that the approach is mechanistic. By contrast, the evolutionary view describes the manifest results of disorder. The issue is whether or not these new insights into the behavior of natural systems have relevance to economics, especially when studying turbulent periods and long-term development.

Certainly, it could be argued that progress in the natural sciences has applicability to and implications for the social sciences. No doubt, there are several areas of potential application. However, it seems to this author that too much optimism is of negative value. One trivial note will suffice here: In the social sciences and especially in economics, the mechanistic (Thoben, 1982) and the energetic (Mirowsky, 1985) analogies have been important. The fact that mechanistic explanation has been questioned in the natural sciences does not, of course, mean that it must be questioned in economics. Analogies do not refer directly to reality, they are only mental constructs. Progress in describing complex systems, for example, with the aid of catastrophe theory, changes the nature of theories, not necessarily because our view of reality has changed but because of progress in our methods.

THE EVOLUTION VERSUS THE NEOCLASSICAL PARADIGM

When comparing the evolutionary paradigm and the core of the neoclassical economics (the *homo oeconomicus* concept), the difference is obvious. It is not possible here to give a precise definition; instead we shall simply illustrate by means of the following pairwise list of attributes associated with the description of change. The first word in each pair refers to neoclassical economics; the second to the evolutionary view:

Table 1

PROCESS/CHANGE

Neoclassical		Evolutionary
deterministic	vs.	stochastic
reversible	vs.	irreversible
controllable	vs.	noncontrollable
predictable	vs.	explainable
(ex-ante)		(ex-post)
quantitative	vs.	qualitative
gradual	vs.	revolutionary
homogeneous	vs.	heterogeneous
linear	vs.	circular

One basic difference lies in the perception of time. The neoclassical view is in a sense timeless (for the so called Cambridge criticism see Cohen, 1984 and Parrinello, 1984). Another related difference is individualism (mainly methodological) of economics in contrast to the holistic view of the evolutionary paradigm. A further important difference is in the assumptive role assigned to the researcher. In economics, the objectivity and passivity (externally) of the researcher is emphasized, while the evolutionary view gives special weight to the awareness of subjectivity and involvement. The latter view emphasizes the difficulty of separating subject and object, reflecting the ideas of relativity theory (see Allen, 1984 and Prigogine and Stengers, 1984). It is argued that nowadays the researcher could, and even should, be responsible for "world-making" in a concrete sense. "We are not the helpless subject of evolution—we are evolution." (Jantsch, 1980, 8). This slightly paradoxical view among others has been strongly opposed by Goldsmith (1982, 239).

Possibly the most controversial difference between the two approaches is the question of free will in a bifurcation regime. According to the evolutionary view, there are better possibilities (freedom) under a bifurcation regime for small actions to follow free will and affect the world than under "normal conditions." This is slightly paradoxical claim for a view which at the same time emphasizes non-controllability and stochasticity of evolution. However, paradoxes are essential ingredients of the new paradigm when it explains evolution through contradictions.

FREE WILL AND DISORDER

As already emphasized, there is as yet no complete theoretical approach in neoclassical economics which could connect theories of individual action (individualism) with more holistic theories embracing "collectivist" concepts such as social norms. We conclude this paper with a few words on free will and disorder. The examples above give reason to be critical of the optimism shown by some representatives of the sciences of evolution: crises and disturbances do not necessarily mean more freedom, especially if freedom is not realized/perceived. Freedom of will when the "will" is not autonomous is a very problematic abstraction.[4]

Laszlo (1985, 19) argues that the determinism of development vanishes when a major perturbation challenges the basic structures of a living and dynamic system. This argument could be a self-prophecy if researchers could convincingly show that necessities are not real." This is partly a political question: How to emphasize the "subject-nature" of a man? How to separate illusions from reality? However, it should be remembered that as a description "free man" could be false when analyzing turbulent societies.

The nature of conflict between voluntaristic and deterministic theories (and voluntarism and determinism) is of utmost importance when trying to evaluate the significance of the complexity paradigm for economic theory. Obviously, neither free will nor system (or structure) alone "determines" behavior. It is possible that at the collective (macro) level disorder means freedom to change the system, as Laszlo (1985) argues, but at micro-level structures and routines followed by macro disturbances could even decrease freedom in everyday activities.

PROPOSITIONS CONCERNING THE LINKING OF THE EVOLUTIONARY PARADIGM TO NEOCLASSICAL CONSUMER ECONOMICS

If the evolutionary view the following propositions concerning the description of an economic agent (cf. figure 1 in appendix) should be considered: *In condition of structural instability* (bifurcation period)—

1) *The inner structure* (hierarchy of preferences and reasoning; normally excluded by the assumption of narrow rationality and given preferences (consumer theory) or given technology (theory of the firm):
   a) *changes;* and/or
   b) *interacts;*
   c) and the result is the *unpredictable emergence of new structures and functions* (inner feedback mechanism; micro-micro connections).

Structures change partly as a result of exogenous shocks and partly endogenously as a result of interaction between parts. The interaction between parts implies the property that the whole is more than the sum of its parts (see Bunge, 1977, 40). By "unpredictable" we refer mainly to epistemological complexity and to the present state of neoclassical economics. Possibly unpredictability belongs to the ontological category of the emergence of unpredictable structures.

The assumption that preferences are given when an economic agent faces a new unexpected environment is a very bold one. There are situations, such as that of Sour Grapes described by Elster (1983) where, when opportunities or resources become more limited, preferences also change or adapt.

2) *Outer structure* (normally excluded by the *ceteris paribus* assumption and independency of exogenous variables; institutional and social black box):

a) *changes*; and/or
b) *interacts* (seemingly autonomous parts of the macrostructure)
c) and the result is the *unpredictable emergence of new structures and functions* (macro-macro connections).

Under structural instability, seemingly autonomous areas (e.g. the economy, culture, politics) lose part of their "independent" status. This could be exemplified by the politics of economic decline (see Alt, 1979). Analytic distinctions, like economics and politics, lose a part of their foundation. Just how much they lose is an open question. Is it worth losing an analytical instrument? In a stable environment and in the short run the independence assumption will suffice.

3) *Inner and outer structures:*
a) *interact;*
b) as a result *unpredictable and unintentional structures and functions arise* (micro-macro connections).

It is important to emphasize that, even though under structural instability development, both at micro- and macro-level, is often unpredictable and unintentional, there is nevertheless a certain logic. For example, disasters such as natural catastrophes do not often give rise to social chaos (for disasters, see Quarantelli, 1978). "Disorder" has its own logic.

Frequently it is a question of the framework in terms of which we analyze the problem. Social psychologists have defined "rules of disorder" by studying children (Marsh et al., 1978). What is disorder from the point of view of adults may well be natural and orderly from the viewpoint of children. Anthropologists have also stressed that, for example, rationality and irrationality are highly context-bound (see Hollis and Lukes, 1983).

In discussing endogenous preferences three separate feedback mechanisms to microstructure and preferences can be studied. The first was macro or outer feedback: microbehavior, macroconsequences, microstructure, microbehavior, etc. (general sociological theory). The second is micro or inner feedback: reasoning, actions, micro-level, consequences, reasoning, etc. (i.e. lifestyle, learning, habits and customs). In addition, there is a third source of preference change: internal structure (psyche etc.) could change, especially in the bifurcation area.

The explanation of consumption patterns when a complex interconnectedness of macro- and microlevels prevails often leads to counterintuitive results—anomalies from the viewpoint of standard economics. It might be useful to explicate phenomena by studying both internal and external reasons for change together. In self-organizing systems internal structure (preferences) has a logic of its own, and at the same time structure also responds to changes in external factors, i.e. in the environment. The joint presence of these dynamics implies a counter-intuitive development in bifurcation periods. It is possible that macro-level disorder (suboptimality, counter-finality, unpredictability) may arise from micro-level order (rationality, predictability)—and vice versa.

In this connection it is interesting to note that Becker (1962) argued that market rationality (i.e. predictability) is consistent with household irrationality. Likewise, the macro-level disorder—economic contraction—could lead to a large degree of order and predictability at the micro-level. Social and psychological structures neglected in standard economics could respond in a stability-maintaining way. Ideological response at the macro-level and defensive response at the micro-level are examples of such possibly stability-maintaining factors.

Clearly the biggest problems stem from the multi-disciplinary nature of the issues. The use of concepts like stability, order, predictability and emergence calls for further analysis. However, to stimulate research on complex economic phenomena requires the construction of preliminary frameworks as a basis on which qualitative development and the suitability of different approaches can be evaluated.

FURTHER PROPOSITIONS

We conclude this paper by presenting three more propositions (which follow partly as consequences of propositions 1–3):

4) Under structural instability, the role of directional time assumes more importance. Symmetric and reversible, i.e. timeless, processes (descriptions) no longer suffice.

5) Under structural instability, simultaneous relations between structures and functions are emphasized. Functions and structures respond to changes in structures and functions.

6) The greater the instability (change) and the faster the rate of change, the more likely it is that propositions 1–5 are of importance to economic analysis. Eventually more emphasis will have to be given to the formation of new structures and conscious and unconscious processes of self-organization. The "emergence of the unexpected"

implies less emphasis on *ex-ante* analysis (predictions) and more on *ex-post* descriptions (explanation).

APPENDIX

A MODEL OF HOMO OECONOMICUS

Substantively Rational Homo Oeconomicus

By *homo oeconomicus* we mean a classical concept used to illustrate an economic agent. It rests on two assumptions (see, for example, Simon, 1976):

(1) The economic agent has a particular *goal* (for example, utility maximization);
(2) The economic agent is substantively *rational*; i.e. he aims at given goals, *whatever* they are.

Simon constrasted substantive rationality with procedural rationality. The latter is based on the findings of modern psychology and is related to deliberation, learning procedures and cognitive processes which are unknown in substantive rationality. It is fair to say that these factors are also partly included and acknowledged in the modern theory of choice (Day and Groves, 1975; Tversky and Kahneman, 1974). However, much modern demand analysis and especially empirical consumer research still lacks these theoretical components.

So far, the *homo oeconomicus* construct implies only the separate existence of means and ends and an internal consistency assumption (narrow rationality).[2] It is important to note that there are several interpretations of *homo oeconomicus* concerning its realism and pragmatic value. For example, Machlup (1978) emphasizes that the "ideal type" assumptions are meant only for studying the results of choices, not for illustrating the choice situation of individuals. However, this is not the view adopted here. On the contrary, one of the main claims we shall put forward is that in order to understand the frequently counterintuitive results of choices it is necessary to study also the internal "structure" of decision-makers.

To attribute psychologism to a neoclassical programme is, as Latsis (1976, 17) pointed out, a fundamental misunderstanding. He argued that the development of marginalism in the 19th century "gave rise to the research programme of situational determinism whose central characteristics are the autonomy of economic decision-making and the deliberate exclusion of the decision-maker's inner environment from explanations of economic behavior.''

Environmentally and Situationally Determined Homo Oeconomicus

Homo oeconomicus is a part of the core[3] of the neoclassical research programme. Latsis (1976, 22) gave the following simplified illustration of its hard core:

(i) Decision-makers have correct knowledge of the relevant features of their economic situation;
(ii) Decision-makers prefer the best available alternative given their knowledge of their situation and of the means at their disposal;
(iii) Given (i) and (ii), situations generate their internal "logic" and decision-makers react appropriately to the logic of their situation;
(iv) Economic units and structures display stable, coordinated behavior.

According to Latsis, the libertarian-rationalistic model of choice is paradoxical:

Decision-makers have only single-exit situations. There is no room for discretion. It is only the logic of the situation, not the individual, which defines behavior (Latsis, 1976). Modern economics excludes not only the decision-makers inner "environment"-psyche, but also the social environment ("outer" structure).

Latsis emphasized the nature of given environments. Inner and outer environments of economic agents are predetermined and constant. Only so-called economic and quantifiable variables, prices and incomes, vary. Information is given (i), the logic of reasoning and rules are given (iii), and the outer environment is given and stable. These assumptions sometimes referred to as the *ceteris absentibus* assumptions, or more frequently the *ceteris paribus* assumptions, imply that problems could arise in economic theory when trying to *understand* structural instability, structure in change. Perhaps even a new language (or analytical apparatus) is needed if complex structures and qualitative change is to be studied.

In other words, when studying complex interconnected phenomena within a neoclassical framework, there are two problematic assumptions. The first is the fixed inner (cognitive) environment, a black box almost without any internal structure. The second is the givenness of the outer environment. All the institutions, norms, rules, laws, etc. are ignored.

Even if their existence is acknowledged the feedback mechanisms (for example, chain: choice (t = 1)–institution (t = 2)–choice (t = 3); t = time) must be neglected for analytical reasons.

We shall argue, following among other Hollis (1983) and Field (1984), that in microeconomics several implicit *ceteris paribus* assumptions and feedback mechanisms should be taken into account explicitly if we are to *understand* the complex development of the economy and society, especially *transition periods* and *evolution in the long run*. The structural variables (parameters), the role of which often emerges when they are threatened by disturbances, consists of several institutional arrangements, norms, institution and human predispositions.

However, it seems that to study disorder is a decidedly more profound task than merely extending the analysis to unexplored areas or increasing the number of variables. Those concepts which normally enable economists to produce a unified picture of reality must also be systematically analyzed.

An Extended Model

For dynamization purposes it is useful here to define the inner and outer environment of an economic agent as *structures,* in accordance with the *structural-functional* analysis by Chase (1979, 1985). A structure consists of those uniformities or patterns which can be discerned or alleged to exist in the phenomena studied. Structural entities can be classified as:
a) historical (mechanistic),
b) non-human biological, or
c) human biological and social.
      Here we refer mainly to the last, i.e. to social entities.

A function is a condition or state of affairs that results from the operation of the relevant structural unit. In addition, contrary to Chase (1979) function refers also more generally to the nature of the processes, the type of reactions and reaction kinetics (Jantsch, 1981, 67).

It is argued in this paper that evolution can be seen as a continuous interplay between structures and functions. Neither is predominant—in contrast to the view of structuralism or functionalism. In economics "structuralist" ideas dominate as long as there are inner and outer structures which dominate behavior. No simultaneous relationship is allowed between functions and structures in spite of the fact that, for example, Schumpeter (1947) argued that to understand change it is necessary to examine such a relationship. In this paper we describe an economic agent who exhibits the interplay between structures and functions.

An agent's control over the consequences of his acts decreases the more aggregated is the level under study. This explains why the inner feedback and outer feedback differ in their nature. It should be recalled that just as "mind" and "body" are two abstractions from one reality—the person—so, too, are "individual" and "society" two abstractions from one reality—the "social individual." This issue is remarkably well-illustrated in Steedman's paper (1980), which concentrates on problems caused for welfare analysis by non-autonomous preferences.

The extended model of *homo oeconomicus* consists of two structures and two functions and several intertemporal feedbacks (in figure 1 only feedbacks influencing individual preferences are stressed).

Figure 1 illustrates explicitly some potential extensions needed in economic analysis if the evolutionary view is to be included. In phase 1 $(t = 1)$, the typical neoclassical explanation is represented by the area enclosed by the dotted line. From the inner structure (i.e. *homo oeconomicus* with preferences), micro-action (consumption) is "deduced." The macrofunction is a result of the actions of micro-agents. It is important to notice that aggregation is "systemic." Interaction between single agents gives rise to a result which is not pure "summation," as already emphasized by, for example, Keynes (1937) in his study of stock markets. Both Schelling (1978b) and Elster (1983b) have argued convincingly that self-serving behavior could lead to unexpected results. Elster has emphasized that certain results, usually assumed to be intentionally generated, are in fact only by-products.

In rationalistic-individualistic analysis the outer and inner structure is excluded from explicit analysis by *ceteris paribus* assumptions. The "direction" is from private needs to general (social) action. As is well known, in functionalistic (Parsonian) explanation the analysis proceeds in the opposite direction: the focus is primarily on so-called system "needs." Neither tradition is sufficient when studying the instability of structures and functions.

General equilibrium analysis, a fundamental part of modern economics, studies mainly relationships between micro- and macro-functions. The only institutions that exist in the economy are markets of the competitive type (Schotter, 1981). All information in the economy must be transmitted through prices formed in such markets. Agents act parametrically and in isolation. It should be remembered that, in spite of the problems of equilibrium concept, "general equilibrium" is one possible and highly sophisticated way to study complex situations created by market agents.

For the sake of abstraction, neoclassical economics usually assumes that structures are almost stable. (For the necessity of fixing goals in

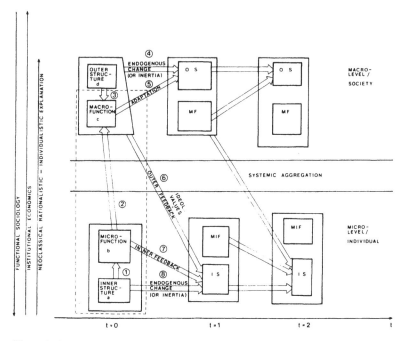

Figure 1. An extended model of homo oeconomicus.

a) *Inner structure:* Preferences, logic (reasoning) and more generally the rationality postulate (= *homo oeconomicus*)—entity. The inner structure leads to (arrow 1):

b) *Microfunction:* Private action determined by the inner structure and "economic" variables (income and prices) (for example, consumption activity).

The microfunctions lead to (arrow 2):

c) Macrofunction: Aggregate behavior (followed by); systemic aggregation includes potential interaction between agents (for example, total consumption). Macrofunctions are a result of microfunctions and the outer structure (arrow 3):

d) Outer structure: Language, constitutive and regulative rules, institutions, norms, etc., which restrict and direct functioning at macro- (also micro-) level (= normally *ceteris paribus*).

optimization models, see Prigogine and Stengers, 1984, p. 207 or Cohen and Axelrod, 1984; for the relative stability of economical structures, see Harsanyi, 1960.) Obviously, in the evolutionary view neither the outer (macro) structure nor the inner (micro) structures can be assumed as given.

Structures have "histories" of their own and they are constantly changing. Here we shall give one, very tentative illustration of how

changes, especially in the inner structure (preferences), might be described (cf. Chase, 1979, 1985). Upward, downward and horizontal links
between the "boxes" in figure 1 illustrate obvious sources of structural
change. In this framework trivial sources of structural change (with one
connection) are as follows:

Endogenous Structural Changes

a) Usually the *inertia of the outer structures* (slow variable) implies
fixed structural entities; *interaction between parts* (cf. polity & economy) could lead to qualitative change, especially near bifurcation
points (case 4).

b) Macro-events (functions) force the structure to respond (functionally
or dysfunctionally); *i.e. structures adapt* (case 5). For example, the
emergence of property rights could be seen as a structural adaptation
when unlimited resources become limited. Ullman-Margalit (1978)
and Schotter (1981) give interesting examples of the emergence of
institutions as solutions to different game-like situations.

c) *Outer feedback;* the inner structure adapts to macro-events and the
outer structure (case 6). These, in turn, are results of micro-activities.
Development of society implies changes in the inner structure. For
example, industrialization—or urbanization—has created new ways
of life (see, for example, F. Hirsch, 1982 or Riesman, 1950). The
outer feedback has not been a major area of research for "mainstream" economics. Haavelmo (1971) is a notable exception. His
concern about implications of rising level of aspirations with increasing consumption ("appetite-arousing cake") are of great significance
for welfare analysis, even if not widely known. Evidently, preferences
are not fixed but rather the results of possibilities (Elster, 1983b) and,
more generally, of social development.

d) *Inner feedback;* a single agent, when making decisions, creates or
limits future decisions (case 7). For example, the chain of "reasoning, action, consequences, reasoning" operates so that decisions are
also functions of the previous chain. If memory or expectations are
included, the complexity of the "inner feedback" mechanism will be
more obvious. There is a conservative lag. Surprises also affect
preferences if the world is too complex for the decision-maker to develop a correctly specified model of the environment (Cohen and Axelrod, 1983). In consumer theory, one version of the inner feedback

mechanism is included in so-called habit formation models (see Pollak, 1976), where the consumption of previous periods (= microfunctions) is assumed to affect this period's consumption (or preferences).

e) *(Endogenous) change (potential) of the inner structure: Inertia of the inner structure* is assumed to be so strong in economics that a fixed structure can be assumed (case 8). Usually this assumption is called the assumption of *fixed preferences*. The reality of this assumption is most threatened when outer feedback (c) becomes stronger (for example, under social transformation), or when the inner feedback mechanism (d) works efficiently (for example, habit information). A third possibility is that the structure itself is near its unstable bifurcation area (for example, in the turmoil of passions). In that case, even a small change in the outer structure could cause a qualitative change in the inner structure and in the action pursued. It is clear that these three sources (c,d,e) of preference change work together and that indirect links between them could be very complicated.

The framework presented in figure 1 attempts to extend the traditional neoclassical model of an economic agent to the principles of the complexity view. In this framework, the *homo oeconomicus* idealization is only one possible specification (upward arrows; dotted line). Complexity ideas can be illustrated by the inclusion of diagonal and horizontal arrows. Their inclusion implies more emphasis on feedback mechanisms and the irreversibility of evolution. Unfortunately, the increase in realism implies a decrease in analytical elegance.

NOTES

1. The unavoidable eclecticism of the paper could make it difficult to obtain a clear picture of the problem at hand. Our examination seeks to apply very general ideas of the complexity view to economics. To follow in this "spirit" is, of course, an exercise which is biased by preconceived attitudes toward economics. One example will suffice to illustrate this point: the author is much inspired (possibly biased) by social scientists like Jon Elster and Amartya Sen, both of whom represent the multidisciplinary tradition of science with a wide area of interests in social practice.

2. In economic theory in the last half century reliance on pure internal consistency conditions has increased significantly, starting with Samuelson's use of "the weak axiom of revealed preferences" (Sen, 1984, p. 9). However, it seems to the present author that there should be at least some further assumptions concerning the existence of means and ends for descriptive purposes. By contrast, the classical assumptions of perfect information or egoism could be omitted when illustrating the core of modern consumer theory. Note that ideas about rational choice have been exported, for example, to political the-

ory. The analysis of linking the neoclassical model and the complexity view is also relevant to political theory where political interests (cf. preferences) are given (Schwartz et al. 1984).

3. In Lakatos' terminology every science is a research programme of a special kind, and a core of a research programme refers to the fundamental statements or axioms on which the research programme is built. These axioms are regarded as irrefutable by the supporters of the programme and are not directly tested. It has been argued that neoclassical economics could be evaluated as a research programme. The discussion of growth of knowledge, paradigms and research programmes (cf. Fulton, 1984) has been spirited and it has been questioned whether it is at all possible to analyze economics within the framework developed in the natural sciences. Here, however, we have taken the Lakatosian concept of core for granted and the well-known article of Latsis (1976) as point of departure.

4. Contraction could reduce alternatives but at the same time ideology and cognitive processess like "sour grapes" could restrict the area of volition and desires. Subjectively defined, well-being could remain unaffected. According to Elster's (1983b) definition one essential dimension of freedom and welfare is the autonomy of wants. Autonomity should be emphasized because it is the constitutive nature of "Man." As Frankfurt (1973, 7) stresses: "No animal other than man. . . . appears to have the capacity for reflective self-evaluation that is manifested in the formation of second-order desires." Second-order desires (or preferences) means the ability of Man to want to have certain desires and motives. This is of utmost importance when analyzing preferences: there exists higher order preferences if autonomy (or freedom) is taken seriously. However, there is one problem: infinite regress (Elster, 1979, 1983b). There are infinitely many "steps upward" if one is trying to reach the autonomous level of free will. The process of higher and higher preferences is never-ending, yet the process should ultimately come to a halt at some level. In this sense freedom is impossible. As Elster argues, the problem of preference formation and endogenous preference change is the greatest obstacle to complete freedom. (Elster, 1979, 162).

REFERENCES

Albin, P. S.—Gottinger, H. W. (1983) Structure and Complexity in Economic and Social Systems, *Mathematical Social Sciences* 5/1983, 253–268.

Allen, P. M. (1984) *Toward a New Science of Complex Systems*, Universite Libre de Bruxelles, Draft version for the Club of Rome, July 1984.

Alt, J. E. (1979) *The Politics of Economic Decline: Economic Management and Political Behavior in Britain Since 1964*, Cambridge Univ. Press, Cambridge 1979.

Becker, G. S. (1962) Irrational Behavior and Economic Theory, *The Journal of Political Economy*, Vol. IXX, February 1962, 1–13.

Benhabib, J.—Day, R. H. (1981) Rational Choice and Erratic Behavior, *Review of Economic Studies* 1981, 459–471.

Boulding, K. (1978) *Ecodynamics: A New Theory of Societal Evolution*, Sage, Beverly Hills, 1978.

Bunge, M. (1977) *Treatise on Basic Philosophy: The Furniture of the World*, Volume 3, Ontology I, Dordrecht, Reidel Publishing, 1977.

Bunge, M. (1979) *Treatise on Basic Philosophy: A World of Systems*, Volume 4, Ontology II, Dordrecht, Reidel Publishing, 1979.

Chase, R. (1979) Structural-Functional Dynamics in the Analysis of The Socio-Economic Systems: "Development of the Approach to Understanding the Process of Systemic Change", *American Journal of Economic and Sociology*, July 1979, Vol. 38, No. 3, 293–305.

Chase, R. (1985) A Theory of Socioeconomic Change: Entropic Processes, Technology, and Evolutionary Development, *Journal of Economic Issues*, No. 4, 1985, 797–823.

Cohen, A. (1984) The Methodological resolution of the Cambridge controversies, *Journal of Post Keynesian Economics*, Summer 1984, 614–629.

Cohen, M.—Axelrod, R. (1984) Coping with Complexity: the Adaptive Value of Changing Utility, *American Economic Review* 1/1984, 30–43.

Crosby, F. (1979) Relative Deprivation Revisited: A Response to Miller, Bolce, and Halligan, *The American Political Science Review*, Vol. 73, 1979, 103–112.

Daly, H. (ed.) (1980) *Economics, Ecology, Ethics: Essays toward a steadystate Economy*, W. H. Freeman and Company, San Francisco 1980.

Day, R. H. (1982) *Dynamical systems Theory and Complicated Economic Behavior*, Modelling research group working paper 8215, Department of Economics, University of Southern California, Los Angeles 1982.

Day, R. H.—Groves, T. (1975) *Adaptive Economic Models*, N.Y., Academic Press, 1975.

Dosi, et al. (eds.) (1988) *Technical Change and Economic Theory*, Pinter Publisher, London, 1988.

Duesenberry, J. S. (1949) *Income, Saving, and Consumer Behavior*, Cambridge, Mass., 1949.

Earl, P. (1983) *The Economic Imagination, Towards a Behavioral Analysis of Choice*. N.Y., M. E. Sharpe, Inc., 1983.

Elster, J. (1978) *Logic and Society, Contradictions and Possible Worlds*, Pitman Press Ltd, Bath.

Elster, J. (1979) *Ulysses and the Sirens, Studies in Rationality and Irrationality*, Cambridge, Cambridge University Press, 1979.

Elster, J. (1983a) *Explaining Technical Change*, Oxford University Press, 1983.

Elster, J. (1983b) *Sour Grapes, Studies in the Subversion of Rationality*, Oxford University Press, 1983.

Field, J. (1984) Microeconomics, Norms and Rationality, *Economic Development and Cultural Change*, 1984, 683–711.

Frank, R. H. (1985) The Demand for Unobservable and other Nonpositional Goods, *American Economic Review*, March 1985, 101–115.

Frankfurt, H. G. (1971) Freedom of the Will and the Concept of a Person, *The Journal of Philosophy*, Vol. LXVIII. No. 1, 1971.

Friedman, M. (1962) *Price Theory: A Provisional Text*, Chicago 1962.

Fulton, G. (1984) Research Programmes in Economics, *History of Political Economy*, 1984, 185–204.

Giddens, A. (1979) *Central Problems in Social Theory—Action, Structure and Contradiction in Social Analysis*. (Finnish edition Otava 1984).

Goldsmith, E. (1981) Superscience—Its Mythology and Legitimisation, *The Ecologist* 5/1981.

Gottinger, H. W. (1983) *Coping with Complexity, Perspectives, for Economics, Management and Social Sciences*, D. Reidel Publishing Company, Theory and Decision Library, Volume 33, Dordecht 1983.

Guha, A. S. (1981) *An evolutionary view of economic growth*, Clarendon Press, Oxford, 1981.

Haavelmo (1971) *Some Observations on Welfare and Economic Growth*, Reprint Series No. 72, University of Oslo, Institute of Economics, Oslo 1971.

Harsanyi, J. C. (1960) Explanation and Comparative Dynamics in Social Science, *Behavior Science* 5/1960, 136–145.

Heiner, R. A. (1983) The Origin of Predictable Behavior, *American Economic Review*, Sept. 1983, 560–595.

Hirsch, F. (1977) *Social Limits to Growth*, Routledge & Kegan Paul, London 1977.

Hirsch, H. (1980) *Der Sicherheitsstaat. Das "Modell Deutschland", seine Krise und die neuen sozialen Bewegungen*, Europäische Verlagsanstalt, Frankfurt am Main 1980. (In Finnish, Vastapaino 1983).

Hirschman, A. O. (1982) *Shifting Involvements, Private Interest and Public Action*, Princeton, Princeton University Press, 1982.

Hodgshon, G., (1988) *Economics and Institutions*, Polity Press, Cambridge, 1988.

Hollis, M. (1983) Rational Preferences, *The Philosophical Forum* Vol. XIV, No. 3–4, 1983, 246–262.

Hollis, M.—Lukes, S. (eds.) (1982) *Rationality and Relativism*, Basil Blackwell, Oxford, 1982.

Jantsch, E. (1981) Autopoiesis: A Central Aspect of Dissipative Self-Organization in Autopoeisis: *A Theory of Living Organization* (ed. M. Zeleny), North Holland, New York 1981.

Jantsch, E. (1980) *The Self-organizing Universe*, Pergamon Press, Oxford 1980.

Kahn, A. E. (1966) The Tyranny of Small Decisions: Market Failures, Imperfections and the Limits of Economics, *Kyklos* (1966).

Karsten, S. (1983) Dialectics, Functionalism and Structuralism in Economic Thought, *American Journal of Economics and Sociology*, April 1983, Vol. 42, No. 2, 179–192.

Keynes, J. M. (1937) The General Theory of Employment, *Quarterly Journal of Economics*, Feb. 1937, 51(2), 209–223.

Klee, R. L. (1984) Micro-Determinism and Concepts of Emergency, *Philosophy of Science*, 51(1984), 44–63.

Latsis, S. J. (ed.) (1976) *Method and Appraisal in Economics*, Cambridge University Press, Cambridge 1976.

Laszlo, E. (1985) The Crucial Epoch, Essential knowledge for living in a world in transformation, *Futures*, Feb. 1985, 2–23.

Malaska, P. (1985) *Organic Growth and Renewal, An outline for post-industrial development*, paper held at the European symposium of the International Council of Social Welfare in Turku June 9–14, 1985.

Marsh, P.—Rosser, E.—Harre, R. (1978) *The Rules of Disorder*, Routledge & Kegan Paul, London 1978.

Machlup, F. (1978) *Methodology of Economics and Other Social Sciences*, Academic Press, 1978.

Machlup, F. (1958) Structure and Structural Change, Weaselwords and Jargon *Zeitschift für Nationalökonomie* Band XVIII, 3, 1958.

Meadows, D.—Richardson, J.—Bruckman, G. (1982) *Groping in the Dark, the First Decade of Global Modelling*, John Wiley & Sons, Chicherster 1982.

Mirowski, P. (1985) Physics and 'The Marginalist' Revolution, *Cambridge Journal of Economics*, 1985, 8, 361–379.

Morgenstern, O. (1966) The Compressibility of Economic Systems and the Problem of Economic Constants, *Zeitschift für Nationalökonomie*, XXVI, 1966, 190–203.

Morgenstern, O.—Thompson, G. (1976) *Mathematical Theory of Expanding and Contracting Economies*, Lexington Books, Toronto 1976.

Nelson, R.—Winter, S. (1982) *An Evolutionary Theory of Economic Change*, Harvard University Press, Cambridge, 1982.

Pantzar, M. (1986) Economic Agent as Changing the Structures and Adapting to the Structures—An Evolutionary View on Consumer Choice, Helsinki School of Economics, Studies B–86, Helsinki, 1986.

Parrinello, S. (1984) Adaptive Preferences and the Theory of Demand, *Journal of Post Keynesian Economics*, Summer 1984, 551–560.

Ploman, E. W. (1984) *Reflections on the State of the Art and the Interest of the United Nations University in the Field of Complexity*, Draft for the Club of Rome, July 1984.

Prigogine, I.—Stengers, I. (1984) *Order Out of Chaos-Man's New Dialogue with Nature*, Heinemann, London 1984.

Pollak, R. A. (1978) Endogenous Tastes in Demand and Welfare Analysis, *American Economic Review*, May 1978, 374–379.

Pollak, R. A. (1976) Interdependent Preferences, *American Economic Review*, June 1976, 309–320.

Quarantelli, E. L. (ed.) (1978) *Disasters, theory and research*, SAGE Studies in International Sociology 13, London 1978.

van Raaij, W.—Eilander, G. (1982) Consumer Economizing Tactics for Ten Product Categories, *Papers on Economic Psycholocy*, Number 26 Erasmus University Rotterdam 1982.

Riesman, D. (1950) *The Lonely Crowd*, Yale University Press, London 1950.

Roeding, W. (1977) *A New Approach to Modelling Some Economic Problems*, 1977, in Lecture Notes in Economics and Mathematical Systems (eds. Beckman & Kunzi) 144, Mathematical Economics and Game Theory, Essays in Honor of Oscar Morgenstern, Springer Verlag 1977.

Schelling, T. C. (1978a) Economics, or the Art of Self-management, *American Economic Review*, May 1978, 290–294.

Schelling, T. C. (1978b) *Micromotives and Macrobehavior*, W. W. Norton & Company, New York 1978.

Schotter, A. (1977) *Economically efficient and politically sustainable economic contraction*, in Lecture Notes in Economics and Mathematical Systems (eds. Beckman & Kunzi) 144, Mathematical Economic and Game Theory, Essays in Honor of Oscar Morgenstern, Springer Verlag 1977.

Schotter, A. (1981) The Economic Theory of Social Institutions, Cambridge University Press, Cambridge 1981.

Shumpeter, J. (1947) The Creative Response in Economic History, *Journal of Economic History*, Nov. 1947, 149–159.

Schwartz, M.—Thompson, M. (1984) *Beyond the politics of interest*, Working Paper, August 1984, International Institute for Applied Systems Analysis, Laxenburg, Austria, 1984.

Sen, A. (1984) Consistency, *Presidential lecture at the meeting of the Econometric Society*, Madrid 1984.

Sen, A. (1976) Rational Fools: A critique of the behavioral foundations of economic theory, *Philosophy and Public Affairs*, 6/1976–77, 317–44.

Shama, A. (1980) *Marketing in a Slow-growth Economy,* The Impact of Stagflation on Consumer Psychology, Praeger 1980.

Simon, H. (1976) *From Substantive to Procedural Rationality,* in Latsis (ed.) 1976.

Spiefhof, A. (1952) The "Historical" Character of Economic Theories, *Journal of Economic History,* 1952, 131–139.

Steedman, I. (1980) Economic Theory and Intrinsically Non-autonomous Preferences and Beliefs, *Quaderni Fondazione Feltrinelli,* No. 7/8, 1980.

Stigler, G.—Becker, G. (1977) De Gustibus non est Disputandum, *American Economic Review,* 1977, 76–90.

Thoben, H. (1982) Mechanistic and Organistic Analogies in Economics Reconsidered, *Kyklos,* Vol. 35, 1982, 292–306.

Tversky, A.—Kahneman, D. (1974) Judgement under Uncertainty, *Science,* Sept., Vol. 185, 1974, 1124–31.

Ullman-Margalit, E. (1978) *The Emergence of Norms,* Oxford University Press, New York, 1978.

Uusitalo, L. (1979) *Consumption style and Way of Life,* Acta Academica Oecomicae, A:27, Helsinki School of economics, 1979.

Veblen, T. (1904) *The Theory of Business Enterprise,* Social Science Classics Series, Transaction Books, New Brunswick, 1978.

von Weizsäcker, C. C. (1977) Notes on Endogenous Change of Tastes, *Journal of Economic Theory,* Dec. 1971, 345–372.

Woodward, S. N. (1982) The Myth of Turbulence, *Futures,* August 1982, 266–279.

Zeeman, E. C. (1977) *Catastrophe Theory,* Selected papers, 1972–1977, Addison-Wesley Publishing Company, London 1977.

Chapter 9

Cultural Evolution: Social Shifts and Phase Changes

RIANE EISLER

Editor's Introduction: It is entirely proper that the closing chapter in this wide-ranging volume of transdisciplinary studies should deal with what is perhaps the most humanly meaningful yet also the hardest to understand issue of all: humanity's cultural evolution. Eisler approaches its problems from a perspective that is different from, yet complementary to, that of Malaska. She considers the evolutionary stages in humanity's nonlinear and discontinuous historical development in reference to the psychological factors that influence, and perhaps even determine, the overall characteristics of a society. To uncover them, Eisler looks at the stages of sociocultural evolution from the "gender-holistic" perspective. In this perspective the relations between the sexes within a social structure are the "figure" that stand out against the overall characteristics of society as "background."

It turns out to be fruitful to re-examine the overall patterns of societal development in terms of the most fundamental of all human relations, that between the sexes. In so doing we can appreciate that in each stage of the evolutionary progression social relations give rise to power structures, and power structures define the overall conduct of society. The psychological factors that underlie social relations are those that characterize the male and the female psyche—values that make for partnership when the female elements come to the fore, and domination when the male elements have the upper hand. These factors seem obvious when we are confronted with them, yet we seldom meet up with their systematic analysis. Perhaps, cultural history has been written too persistently from a male perspective, where power relations appear as "figure," with relations between the genders as elements merely of the "background."

A complex phenomenon such as cultural evolution can be viewed from various perspectives, and the Gestalt-switch from one to another, though initially difficult, can become easy and productive in time. The perspective projected by Eisler, though novel in the social science literature, is likely to prove especially valuable as we contemplate our options for the future. For, as she—in full agreement with Malaska and Artigiani—points out, we are now at the threshold of a major societal bifurcation. Humanity can break down and enter a path to extinction, or break through to a new and more stable—and perhaps also more humanistic—historical stage.

According to Eisler, in the five-thousand-year epoch that separates us from the Neolithic, values inspired by the male half of humanity have dominated those inspired by its female half. Be that as it may, it is clear that if today's dominant values remain aggression expressed in war and violence, they will bring us to the brink of unprecedented catastrophe. A more partnership-oriented society hallmarked by the valuation of cooperation and egality—between sexes, as well as between classes, states, nations, races, peoples, and cultures—is urgently needed. To fill this need, we need to study the panorama of humanity's cultural evolution in all its manifold perspectives. That provided by Eisler appears to be of exceptional interest.

For the first time in evolution, a species is consciously becoming aware of the possibility of its own imminent extinction. One manifestation of this growing human perception appears to be the resurging scientific interest in constructing better models of evolution.

What can we learn from these new evolutionary models? Even more specifically, how can new models for the study of cultural evolution help us deal with the mounting threat of nuclear holocaust and/or ecocatastrophe?

The emergence of our species marks the start of the co-evolutionary age: the use of our minds and hands to profoundly alter our earthly habitat. From the very beginning, human-devised tools have made us cocreators with nature of our history and our future. And in our time, human technologies—and human culture, which critically affects how we use technology—may determine our future, and whether we even have one.

This paper presents a new model for the study of human culture and human technology. Like other new evolutionary models, it derives from a new scientific paradigm focusing on process. Concerned with how systems emerge and maintain themselves, as well as with shifts from one type of system to another, these models provide a better understanding of the major phase changes that punctuate the evolutionary process that began in the universe ten to twenty billion years ago.[1]

An interesting leitmotif in this rich and multifaceted evolutionary pageant moving from prebiological, to biological, to cultural evolution is a binary principle: the combination at ever higher levels of complexity of pairs of basic, mutually complementary elements. In the formative stages, following the explosive birth of our expanding universe, these binary elements were matter and energy (or in its original form, radiation). During the next great phase change, the emergence of life, nucleic acids and proteins combined in a variety of forms. Then a new set of binary elements required for the survival of heterosexually reproducing species, including our own, was introduced: the female and the male.

Logically, as we move to the level of cultural evolution, how this most fundamental of relationships is organized has important implications for both the structure and function of human social systems. More specifically, the construction of models taking into account the possible ways the relationship between the female and male halves of humanity may be organized should provide useful tools for the analysis of not only our past but our present and future. These models can give us new conceptual maps to help guide our cultural evolution in directions that could preserve human society at a time of increasing societal chaos and lead to higher levels of cultural and individual development.

My own efforts to construct a conceptual map for cultural evolution take into account this fundamental division of humanity into two asymmetrical interrelating halves. Based on a transdisciplinary study of cultural evolution, this new conceptual map focuses on the interaction of two processes. One is a generally unidirectional movement through the five basic phases of cultural development that will be sketched out in this paper. The other is movement between two basic models for social organization, which, by contrast, is multidirectional rather than unidirectional.

SOCIAL SYSTEMS: TWO BASIC MODELS

If we look at human social systems from the perspective of how we structure the relationship between the two fundamental halves of humanity—women and men—and if we then consider the implications of what we find for the totality of our social and ideological system, two basic types or models for social organization become apparent.[2]

The first of these models is a *dominator* model. This model of social relations is primarily based on the organizational principle of *ranking*. This is a way of structuring all social institutions—from the family, religion, and education to science, politics, and economics—so that one type of human is more highly valued than another. Depending on which basic human type—the female or the male—is ranked over the other, this model can take two forms. One would be the ranking of women over men, or a matriarchal or gynocratic form of social organization. The second, which has prevailed over recorded history, is a patriarchal or androcratic social structure based on the ranking of men over women in a domination hierarchy ultimately based on force or the threat of force.[3]

The other model is a *partnership* model, which has a single basic form. Where the dominator model primarily uses ranking as an organizational principle, the primary organizational principle for the partnership model is *linking*. Here neither half of humanity is ranked over the other, with both sexes tending to be valued equally. The distinctive feature of this model is a way of structuring human relations—be they of men and women, or of different races, religions, and nations—in which diversity is *not* equated with inferiority or superiority.

But systems models are by definition abstractions. Before we can see how the dominator and partnership models relate to social shifts and phase changes, we must define them in cross-cultural and historical terms. Specifically, we must see how this primary set of models is embedded within and guides social process in actual social systems.

The Dominator Model

Looking at social organization from a gender-holistic perspective i.e., one that takes into account the data base for both halves of our species), let us examine four societies from different corners of the Earth: 1) an oriental feudal system, the Samurai of Japan; 2) a highly technologically advanced western system, Nazi Germany; 3) a far less technologically advanced African tribal system, the Masai of East Africa; and 4) a religious fundamentalist system in the modern Middle East, Khomeini's Iran. These societies are generally considered to be disconnected phenomena of fundamentally different cultural heritages, value systems, and social organization. But if we re-examine them from the perspective here proposed, we find that underlying their radically different surfaces is a distinctive pattern of structural and ideological commonalities.

Japanese scholars have shown that the Samurai period in Japan followed an earlier period when women, and the "softer" values stereotypically associated with femininity, had a higher social status. The Samurai culture arose from a sharp swing not only toward rigid male dominance, but also toward a rigidly stratified and authoritarian system, in which the samurais or warriors—as well as the fighting of wars—were accorded highest social and ideological value.[4] Similarly, the emergence of Nazi Germany was characterized not only by a shift toward warfare and other forms of institutionalized social violence, such as pogroms and death camps, as well as the rise of fascist totalitarianism, but also by the re-institutionalization of rigid male dominance. At the same time that official Nazi policy proclaimed that the "heroic" male attributes of aggression, dominance and conquest were the highest social virtues, it brought about a marked regression toward rigidly stereotypical male-female socialization. As European historian Claudia Koonz points out, the ideal male portrayed in Nazi ideology was a warrior and the ideal woman was his mother.[5]

Masai socialization likewise hinges on the identification of male identity with dominance and aggression, with its "highest" expression in the role of the warrior. Here men hold all the economic and political power, with the only real status for women—who perform the bulk of the socially productive work, often including even the construction of their own houses—hinging on their ability to give birth to a male.[6] Similarly, the idealization of the male as the "holy" warrior is a prime feature of Khomeini's Iran. Originally expelled by the Shah for inciting a bloody riot against legal reforms giving Iranian women greater social and familial rights, Khomeini upon his return proclaimed the chuddar, the full length dress with which women must shroud themselves, the official flag of his extremely authoritarian fundamentalist religious revolution. And, beginning with the violent takeover of the U.S. Embassy, Khomeini's Iran has also been characterized by another key component—a high degree of institutionalized social violence.[7]

The common feature of these seemingly diverse social systems is that they orient primarily to the same basic model of social organization: the androcratic version of the dominator model. Like earlier examples of this model, such as the theocracies of ancient Sumer and Judaea and Aryans, Dorians, and other Indo-European tribes, such societies have a characteristic social and ideological configuration. This systems configuration is male-dominance, a generally hierarchic and authoritarian social organization, and a high degree of institutionalized social violence, including warfare.

This configuration is characteristic of androcratic social organization regardless of the period in history or the technological level. Moreover, the more closely a social system approximates the male dominant form of the dominator model, the more highly such stereotypically "masculine" values as aggression, dominance and conquest, which are required to maintain this system, tend to be valued. In these rigidly male dominant societies, qualities like caring, compassion, and peacefulness may be given lip service, but in operational fact they are generally considered appropriate only for women and "weak" or "effeminate" men.[8]

The Partnership Model

By contrast, in societies that most closely approximate the partnership model, stereotypically "feminine" values are not devalued. Consequently, the more closely a social system approximates the model of social organization where the male half of humanity is not ranked over the female half, the greater the probability that it will be more peaceful as well as less generally hierarchic and authoritarian.[9]

Until recently, societies patterned on a partnership model were generally believed to exist only at the most technologically primitive level,
among tribes such as the BaMbuti and !Kung.[10] In the 19th century,
archeologists and historians of myth did find evidence indicating there
were more advanced prehistoric societies that were not androcratic or
patriarchal. But they assumed that, not being "patriarchal," these societies were "matriarchal"—in other words, the second form or reverse
variation of the dominator model.[11] Then, when later evidence did not
support the view that women dominated men in these earlier societies, it
was concluded that they must, after all, have been societies where men
dominated women. But more recent archeological findings, as well as a
closer re-examination of ancient myths, indicate that neither conclusion
is valid.

There have always been legends about an earlier, more harmonious
and peaceful age. In our Bible we read about the Garden of Eden, before
a male god decreed that women henceforth be subservient to man. The
Chinese Tao Te Ching recounts a time when the yin or feminine principle was not yet ruled by the male principle or yang. But it was generally
assumed that these were only idyllic fantasies, expressions of the universal human yearning for seemingly impossible goals. Only now, thanks to
new scientific methods of probing our past, the actual facts that lie behind these legends are coming to light. When archeology was still in its
infancy, excavations unearthed the real Troy of Homeric legend. Today
new excavations, coupled with reinterpretations of older digs using more
scientific methods, reveal that these legends about an earlier more
peaceful and harmonious time also derive from folk memories about real
flesh and blood peoples—who organized their societies along very different lines from ours.

Just as in Columbus' time the more popular recognition that the Earth
is not flat made it possible to find an amazing new world that had been
there all the time, our increasing archeological knowledge opens up new
and amazing vistas of our hidden past—and potential future. They reveal
a long period of peace and prosperity when our social, technological,
and cultural evolution moved steadily upward: many thousands of years
when all the basic technologies on which civilization is built were developed in societies that were not male-dominant, violent, and hierarchic.

Specifically, these findings indicate that civilization was not, as we
have been taught, born in Sumer 5000 years ago, but that there were
actually a number of cradles of civilization, all of them thousands of
years older.[12] For example, in Europe, once thought to have been a cultural backwater until the rise of the Hellenic civilization, there is now

evidence of stable Neolithic societies where the arts flourished, where people peacefully tilled the soil, traded, and engaged in crafts, and where there was even a written script predating Sumerian writing by about 2000 years.[13]

In these societies, the characteristic social and ideological configuration or pattern appears to have been basically nonhierarchic or egalitarian: although there were differences in status or wealth, these were not extreme.[14] There are also specific indications that these were *not* male-dominant societies; women were priestesses, women were craftspeople, and to us most surprising if not shocking, the supreme deity was female rather than male: a Goddess rather than a God. As Gimbutas writes, before Old Europe was overrun by Indo-European hordes and "millennial traditions were truncated," the female was seen as "creative and active," with neither the female nor the male "subordinate to the other."[15] Finally, these were also societies that, wholly contrary to our prevailing view of "human nature," do not appear to have had wars. Throughout literally thousands of digs, there is a general absence of fortifications, as well as a general absence, in their extensive and considerably advanced art, of the glorification of warriors and wars.[16]

For example, in Minoan Crete, in marked contrast to the other high civilizations of the time—which were male-dominated, highly authoritarian, and constantly at war—there are no great statues or reliefs of those who sat on the thrones of Knossos or any of the palaces, nor are there any grandiose scenes of battle or hunting.[17] As reported by Nicolas Platon, the former director of the Acropolis Museum who excavated in Crete for over 30 years, on this island, where "the important part played by women is discernible in every sphere," the "whole of life was pervaded by an ardent faith in the goddess of nature, the source of all creation and harmony. This led to a love of peace, a horror of tyranny, and a respect for the law."[18]

Once we look for the more peaceful, more sexually egalitarian, and generally less hierarchic social configuration characteristic of societies that orient to a partnership, rather than a dominator, model of social organization, we also observe the patterns characterizing an ancient civilization like Minoan Crete recurring in some of today's technologically developed societies. A notable example is Sweden, where in a nation that has not been at war for over a century we see a strong movement toward both sexual equality and generally egalitarian social structure.

Moreover, as we begin to look at the full span of our cultural evolution from the dual perspective here proposed, we see the operation of two major processes or dynamics. One is the unidirectional movement toward greater technological, and thus social, complexity. The other is the multidirectional movement between periods primarily orienting either to a partnership or dominator model of social organization as "attractors" for all social systems.

The interaction of these two movements is the basis for the new conceptual map of cultural evolution sketched briefly in this paper, beginning with a fresh look at the major technological phase changes.

CREATIVITY, TECHNOLOGY, AND PHASE CHANGES IN CULTURAL EVOLUTION

A striking aspect of the evolutionary process is the movement to ever greater levels of complexity, offering progressively more flexibility and choice. But with the emergence of the human species, choice takes on a wholly unprecedented dimension. It becomes not only the ability to *select* among given realities but the capacity to *create* new realities.

In the context of cultural evolution, human creativity can be perhaps most accurately defined in the Biblical sense of creation. It is the capacity, first, to image new realities, and beyond that, to actualize these images through technology. Our most taken for granted realities are the actualization of ideas once entertained in the minds of women and men. This is true of such everyday artifacts as pots, tables, rugs, and clothes, and of more complex inventions such as machines, steamships, missiles, and computers. It is also true of the social inventions that everywhere surround us, our schools, hospitals, and stock exchanges, as well as the complex of economic, educational, governmental, religious, and scientific institutions that so profoundly affect every aspect of our lives.

It is this human *creativity*—our capacity to image and actualize new realities through human and human-made technologies—that at ever more advanced levels of technological development makes us ever more active co-creators in the evolutionary process. But the term *technology* is here used in a larger, once again rather Biblical sense of an active role in creation. In this broader definition, technology is not limited to its conventional meaning of human-made tools or technics. It encompasses, first and foremost, our distinctively *human* technologies—our extremely powerful and creative brains and our uniquely creative hands—with technics or tools viewed as what they in fact are: the extensions of these creative capacities through human inventiveness.[19]

From this larger perspective, the grand theme of our cultural evolution—and of the five phase changes proposed below—is the *expansion* of our creative powers. And a key to these five phases is the broader definition of technology as the instrument of creativity: as human or human-made means to attain human-defined goals.

Based on this view of technology, what follows is a scheme that views the emergence of our species as itself a major technological break-through. From this perspective, within the span of known life on Earth the period since the emergence of our species may be described as the Age of Co-Evolution, cumulatively progressing through five major technological phase changes.

I. The Human Emergence Phase:
The Beginning of Co-Creation

Spanning a period of several million years, this first phase marks the beginning of the co-evolutionary age. It entails a transformation from one living system to another: from hominid to human. And it marks a second fundamental transformation, with profound effects on not only human, but planetary, evolution: from living systems that may at best select among given realities to a new form of life with the capacity to *create* new realities.

In the sense of technology as human or human-made means to achieve human-defined ends, this first phase change is this first major technological expansion. It is the emergence of the two most basic of creative technologies—our human brains and hands that made possible the beginnings of human culture. And it is also during the first phase that we see the emergence of the first human-made tools and artifacts, including that most fundamental of human conceptual tools, language. This initiated a co-evolutionary process that gradually, and then at ever more accelerated rates, profoundly altered our terrestrial, and to some extent now also our extraterrestrial, environment.[20]

It is in this initial phase that the two basic forms of human social systems first make their appearance. While the traditional assumption has been that "man, the hunter" was the sole protagonist of our early cultural evolution, it is now evident that "woman, the gatherer" also played a major part in this process.[21] As indicated earlier, there is evidence that in the first formative stages of technological development there was a tendency for human society to move toward a partnership model of social organization.

Bearing this out are the compelling examples noted earlier of two of the most technologically primitive societies surviving into our time, i.e., the BaMbuti of the African Congo and the !Kung of the Kalahari desert.[22] Moreover, in the already much more technologically developed

European Paleolithic, there are strong indications of a gynocentric or women-centered rather than androcentric or man-centered ideology.

Based on his exhaustive analysis of the period, the noted Paleolithic scholar André Leroi-Gourhan stresses how "feminine symbols" occupy the central place in Paleolithic artifacts and art.[23] These kinds of data indicate the need for a revision of the old androcentric view of cultural evolution. A plausible new conceptual map designates a basic tendency in the mainstream of early western cultural evolution toward a partnership rather than a dominator model. This tendency becomes even more evident as we move into the second major phase change, from the gathering and hunting Paleolithic to the agrarian Neolithic.

II. The Agrarian Age: Co-Creation With Macroscopic Organic Matter

The second major phase change in cultural evolution was the first technological milestone in human culture—and the first major step in the alteration of our natural environment. This is the shift from the use of human and human-made technologies for the harvesting of food and other natural resources to the use of technology for our *co-creation* with nature of natural resources.

With this second major technological expansion, the shift into the agrarian age, came the co-creation of a vastly increased and far more reliable supply of food and other organically co-produced means of life support, such as stock breeding. This, in turn, provided the basis for the human creation and/or further refinement of most of the basic technologies on which later civilizations are founded, such as the manufacture of fibers into clothing, the use of clay and wood for housing, the building of boats for transportation and trade, and the smelting of metals for both ornaments and tools. It also made possible the expansion of humanity's great physical and mental powers through the creation of a far more complex system of culture, including intricate systems of government, religion, and arts.

This second phase spans many thousands of years, with its first beginning for western civilization now generally assigned to circa 10,000 B.C.E. (although there are indications that the first use of seeds probably goes back thousands of years earlier).[24] Indeed, it is important to keep in mind that these processes are not merely cumulative, with the breakthroughs of one phase carried over into the next. Rather than consisting of a series of abrupt and discrete developments, each phase change is itself a gradual process where new and isolated nucleations eventually culminate in a qualitative system change.

While the traditional view of the second phase of human cultural evolution has also been androcentric or male-centered, it is clear that this view is not congruent with massive evidence of a complex system of Neolithic religious imagery centering on an all-giving, all-nurturing Great Mother.[25] The seemingly universal veneration in all the major early centers of agriculture—in Asia Minor, Southeast Asia, and Old Europe—of a female deity from whose womb all life is born and into whose womb all life returns at death to be once again reborn, would seem to reflect a social guidance system in which so-called "feminine" values such as nurturance and affiliation were accorded highest social value. As indicated earlier, these societies appear to have been generally peaceful, with an egalitarian social structure based on the partnership of women and men.[26]

But while the original thrust in this formative phase of human culture was toward a partnership model of social organization, the archeological record evidences a fundamental *shift* in the *direction* of our cultural evolution during the second phase.[27] What ensues is first a period of intense physical and social disruption, a chaotic time span lasting several millennia. And when after this period of cultural devastation and loss civilization re-emerges, it is on a radically different course.

The salient characteristic of the period of androcentric ideology and androcratic social organization that follows is *a radical shift in technological emphasis,* required for the purpose of enforcing a hierarchical social organization—a shift that presages the technological and social crises of our time. Previously, the main emphasis had been on technologies to *create:* technologies to sustain and enhance human life, guided by the life-giving image of a Great Mother. The new emphasis is on technologies to *destroy:* technologies to enable men to dominate and conquer, guided by the life-taking image of such seemingly diverse, but essentially same, gods of war as the Greek god Ares and the Hebrew Jehovah. Indeed, now even female deities like Athene and Ishtar, through the age-old process of co-optation, become identified with warfare.

As religious historian E. O. James points out, these deities' identification with warfare was a metamorphosis of the goddesses of old.[28] And, as archeologist Marija Gimbutas notes, this ideological metamorphosis in turn reflected a metamorphosis of the social system.[29]

In most Bronze Age civilizations, with the notable exception of Minoan Crete, the course of Western civilization become profoundly altered. War, or the threat of war, acted as an agent of cultural diffusion. Like a kind of social virus, it spread a dominator model of society—and with it the idealization of the male as the dominator-warrior—to every corner of the ancient world. As a result, societies once patterned around a partnership model were either destroyed or taken over. Or they themselves gradually shifted to a warlike, hierarchic, authoritarian and male-dominant form of social organization.

Nonetheless, even after this fundamental social and ideological transformation, the earlier partnership model of society continued to act as a "periodic attractor" in systems dynamics terms. That is, as we enter recorded history we begin to see oscillations between periods of movement toward a partnership model, which are followed by periods of regression toward a more faithful approximation of the dominator model.

An arresting example from Western history is the shift from early or so-called primitive Christianity to the rise of the institutionalized or "orthodox" church as an arm first of the Roman Empire and then of the Holy Roman Empire. If we look at early Christianity as a move toward a partnership model of society that failed, we find clues to the mystery of how so much bloodshed and cruelty could have been perpetuated in the name of Christian love. The "partnership" ideology was strikingly expressed in Jesus' teachings that there shall be "no master or slave" and "no male or female." Indeed, in early Christian communities such as those described in The Gnostic Gospels, women and men worked in partnership, living and preaching non-violence.[30]

In sharp contrast was the emergence of the "orthodox" Church, which marked a return to a rigidly male-dominated, hierarchic, and violent system. Run from the top by men who vilified women as the "carnal source of all evil," this Church allied itself with Roman authorities to persecute and kill anyone perceived as a threat to the established system.[31] Long after the fall of the Roman Empire, aspiring to create and consolidate a Holy Roman Empire, and still in the name of Christian love, it unleashed the terrible violence and repression of the Holy Crusades and the Inquisition.

If we scan the circa 5,000 years of recorded history looking for the contrasting three-way configurations characteristic of dominator and partnership societies, we begin to find an organizing principle for what otherwise appear to be merely disconnected events without real pattern or internal logic. For example, the extreme misogyny of such rigidly male-dominant systems as the medieval Church is generally perceived as just a historic perculiarity. So also is its fanatic persecution of the Troubadors, who accorded women and "feminine" values high status, and were instrumental in instituting the worship of the Virgin Mary as the Mother of God. But the dominator and partnership models as periodic attractors provide us with new insight into the systems dynamics underlying such hitherto mystifying workings of history. Now these puzzles may be seen as embodying systems maintenance mechanisms for the violent reinstatement of a dominator model of social organization.

Shedding further light on the "orthodox" Church's persecution of seemingly harmless and characteristically non-violent heretics is a revealing but rarely noted feature of both some original or "primitive" Christian sects as well as later "heretics." As Pagels notes, these non-hierarchic and non-violent sects frequently conceptualized the deity as both female and male, for example in the form of both a divine Mother and Father.[32]

As we move into the next technological phase change, we see how, largely due to improved technics of destruction, the historical impact of these neglected systems dynamics becomes progressively greater. When the attraction of the partnership model became even stronger, the androcratic regressions became more violent.

III. The Machine Age: Co-Creation With Macroscopic Nonorganic Matter

The third phase change is the shift from the use of hands and tools for the production of material objects to the *creation* out of nonorganic matter of a new order of technics: machines. Just as the shift into the agrarian age brought the creation of new sources of life support, the shift into the machine age brought the creation of new sources of labor and energy for the infinitely more efficient creation of vastly more diverse and complex material objects. This major technological expansion in turn led to a still greater expansion of human physical and mental powers through the emergence of modern science and the resultant far greater complexity in human ideological and social structures.

The machine age is sometimes equated with the modern or industrial age. But just as the first sporadic use of agrarian technology goes back thousands of years before the date assigned to the agrarian revolution, the construction of simple machines also goes back to an earlier time. Indeed, while it is customary to speak of the roots of the machine age in the first scientific revolution of circa 1300, or at most in the time of the Greek philosopher-scientists of ca. 600 B.C.E., it is clear that earlier ancient civilizations, including Crete, had already begun to take the first steps toward this third great phase change.[33]

To understand what happened during the machine age, it is important to focus on the system maintenance requirements of a dominator system—as distinguished from the system maintenance requirements of the human species.[34] For then we may begin to note two phenomena of great importance for both cultural evolution and species survival (verified by social scientists as diverse as cultural historian Theodore Roszak, social anthropologist June Nash, and social psychologists David McClelland and David Winter). The first is visible in literature and the arts. Regressions to a closer approximation of an androcratic system are characteristically accompanied by more idealization of male violence and domination and, even more specifically, more idealization of male dominance over women. The second is that these ideological changes presage periods of warfare and repression.[35]

In a highly original study of the period in modern western history that preceded both World Wars, Roszak stresses that this was a time when—transcending the conventional ideological distinctions between capitalist, communist, royalist, and fascist—there was a massive re-idealization of "hard" or masculine values.[36] Likewise, in her study of the emergence of the incredibly bloody and tyrannical Aztec empire, Nash shows how this development in Meso-American history was presaged by the refashioning of Aztec mythology to downgrade the influence of formerly powerful life-giving female deities, and the ascension to supreme power of a death-wielding god of war.[37]

Similarly, in his study of shifting themes in modern American and European history, McClelland shows how the decline in importance of "feminine" values, such as "affiliation," and the re-idealization of the violence-oriented complex of "masculine" values he calls "imperial power" presage times of both warfare and repression.[38] And Winter's study indicates that the resurgence of Don Juan type stories, with their theme of conquest—i.e. the rape and degradation of women—is an even more specific predictor of periods of aggressive warfare.[39]

From this perspective it is possible to see how *social shifts* between the partnership and dominator model have profoundly affected the *direction* of cultural evolution. This approach provides new tools for analyzing the totality of cultural evolution in terms of such critical variables as whether it will be peaceful or warlike, liberative or repressive, and egalitarian or authoritarian. It also makes it possible to move toward a view of western history that takes us beyond such conventional ideological labels as Right versus Left.

To illustrate, after the overthrow of tsarist tyranny, there followed a marked movement in the Soviet Union toward equality, or what we would term partnership, between women and men. But culminating with the bloody Stalinist regime—when millions of Kullaks were killed and dispossessed and a reign of terror rivaled only by Hitler was instituted—

a regressive swing toward a new version of a dominator society took place. As historian Sheila Rowbotham and others have documented, at the same time that the Soviet Union became ever more authoritarian and warlike, laws passed in the early days of the revolution giving women reproductive freedom and equal family and social rights with men were either ignored or repealed, as women party leaders considered "too feministic" were purged.[40]

This basic pattern can be seen in other social revolutions, in both modern and pre-modern times. In fact, once we look at recorded history as the stage of periodic shifts toward the partnership, and then back again to the dominator model, much that otherwise seems mysterious, random, and unconnected acquires specific meaning in terms of systems maintenance and systems change.

For instance, we may see that because the prevailing social model during the machine age has been the androcratic dominator model, the technological emphasis has been overwhelmingly on conquest and domination. A recurring theme has been the use of technics for the conquest of nature. These technics have brought many benefits for humanity, but also at great human and natural cost, including the extinction or threatened extinction of many species.

As we see all around us, this is a conquest against which even nature seems to be rebelling. Or so one might read the progressively more dramatic demonstrations of air and water pollution, desertification, deforestation, acid rains, and—as industrial technology continues to be used in conquest of, rather than cooperation with, nature—the ultimate threat of ecocatastrophe.

Industrial technics have also been instrumental in the massive exploitation of the labor of women and men to benefit a few men on top. Against this have also come progressively more rebellions, beginning with the European revolutions of the 18th and 19th centuries and continuing in the 20th century with anti-colonial and anti-racist revolts.

Rebellions against the generally hierarchic and authoritarian character of the dominator model of social organization were, at the height of the machine age in the 19th and 20th centuries, joined by an even more fundamental revolt. This is the feminist movement's challenge to the basic component of the androcratic system: the ranking of the male over the female half of humanity. At the same time, the challenge to the third major component of androcracy, a high degree of institutionalized violence, has also created popular pressure against war, as well as against such forms of institutionalized male violence as wife battering and rape.

It is said that industrial technology has desensitized us. But how do we then account for the lack of empathy and compassion characteristic of pre-industrial dominator societies, such as the Indo-European Aryan invaders, the Huns, or for that matter the Hebrews, Moslems, and Christians who, in so-called ages of faith, tortured and killed their fellow humans with impunity and even zest? A question that might more usefully be asked about the lack of empathy characteristic of the machine age is whether it was in fact the inevitable accompaniment of this major technological phase change, or whether the machine age took the direction it did because it was guided by an androcratic ideology of conquest and domination. This question acquires even greater urgency in the next major phase change: the move from the machine to the electronic/ nuclear age.

IV. The Electronic/Nuclear Age: Co-Creation with Microscopic Nonorganic Matter

The fourth major technological phase change in cultural evolution is the shift that occurred only a few decades ago from the use of human brains to process information to the creation of electronic brains, as well as the co-creation with microscopic nonorganic matter of sources of energy rivaled only by the sun: the shift to the *electronic/nuclear age*.

This technological phase change introduced dramatically effective means of obtaining and analyzing information through computers, as well as technologies of transportation, and above all communication, creating new conditions on our planet. It also introduced exponentially more effective sources of labor, including robots as electronic substitutes for routine functions previously requiring human brain and muscle power.

This major expansion of our physical and mental powers brought with it much greater social complexity, as well as the capacity to analyze complex systems interrelationships. In turn, this has led to the emergence of a new scientific paradigm focusing on interconnections and process.

The fourth phase takes us into a world that is, through human-made technologies, inextricably linked into one interdependent system. We reach a point where our species has become the single most powerful evolutionary force on this planet, culminating our technological expansion from the human technologies of hands and brains to the emergence of human-made technologies as powerful, and in some cases even more powerful, than those available in nature. These technologies include the capacity to create new organs for human bodies, ultimately even the capacity to create life in the laboratory. But they also include an unprecedented power to take, rather than create life, potentially even the power to destroy all life on this planet.

Because of the intensifying appeal of the partnership model as a systems state "attractor," in this phase we also see intensified androcratic systems maintenance pressures. One manifestation is the revival of religious dogmas exhorting a return to "traditional"—i.e. androcratic—values, as well as the return to strong-man rule in both the home and state, a phenomenon occurring in both Eastern and Western cultures.[41] Another manifestation is the re-idealization of "masculine" aggression and conquest as well as the flood of pornographic images depicting the degradation, brutalization, rape, torture, and killing of women—some images hauntingly reminiscent of of the drawings of the torture and burning of witches by the medieval Church—which the works of Winter and McClelland indicate presage the ultimate manifestation of androcratic regression: a new outbreak of war.

But because this next war may be our last, the movement toward the partnership model is also intensifying. This is manifested in the unprecedented global linking of feminist conferences, such as during the First United Nations Decade for Women, with its interrelated goals of equality, development, and peace; in international demonstrations against war involving millions of women and men; in the spread through the human growth and human potential movements of a heightened consciousness of unprecedented possibilities for human actualization rather than destruction; and, as a concrete response to nature's continuing rebellion against androcratic conquest, an international ecology movement—all attempts on a global scale to complete the shift from a dominator to a partnership model of social organization.

V. The Human Actualization or Extinction Phase

We are now approaching a bifurcation point in human cultural evolution. On one side lies the culmination of human creativity on this earth: our breakthrough to the Human Actualization Age. On the other lies the end of our evolutionary journey: the breakdown of human evolution and the possible extinction of our species.

We know from studying the dynamics of phase changes that when a system approaches a critical bifurcation point it may not be possible to predict the course it will take.[42] But it may be possible to predict which factors or interventions will amplify a desired nucleation and which will tend to arrest this movement. In the course of this discussion we have focused on a factor that will determine whether our next phase change will be a breakdown or a breakthrough on a scale beyond all previous human experience; the choice between two basic models of social and ideological organization to guide us.

As we have seen, and today see all around us, the structural or systems maintenance demands of the androcratic dominator model require goals and values emphasizing domination and conquest and the application of technology to assure these ends. As a result, regardless of their level of technological development, societies patterned on this model tend to give highest priority to technologies of destruction—be it in the human form of the "noble warrior" or the human-made form of spears, guns, or nuclear bombs. By contrast, societies that more closely approximate the partnership model give priority to technologies that support and enhance life. One of the lessons of recorded history is that the times of greatest creativity have been periods such as the time of the Troubadors, the Elizabethan Age, and the Renaissance, periods when what G. R. Taylor calls "matrist" (in other words, stereotypically "feminine") values, as well as women, have risen to higher status.[43] This is also the message from prehistory, culminating in the high civilization of ancient Crete, which produced an art that is unique in the annals of civilization.[44]

But there is another message from our prehistory that is just as relevant in our technologically inter-linked globe. This is that a partnership society can only be sustained either in isolation or when the societies it is in contact with are not dominator societies. Futurist phrases such as the urgent need for a new "global ethic" based on "a spirit of truly global cooperation, shaped in free partnership"[45] intuitively capture this fact: that the shift to a different model of social organization must be global. The urgency of these prescriptions for global survival capture still another fact: that, given our present level of technological development, this shift must also be rapid.

Even more striking is the implicit recognition by many futurists that the "drastic changes in the norm stratum"[46]—in our system of values and goals—constitute precisely the shift we have been examining: the shift from a dominator model of society, which must give highest value to the "hard" qualities stereotypically associated with "true" manliness or "masculinity," to a partnership model of social organization where the "softer" set of affiliative-nurturance values stereotypically associated with women and "feminity" are oriented to the generation, maintenance, and enhancement of life. For what they tell us is that we must move from "competition" to "cooperation," from "chronic warfare" to "peaceful coexistence," from "selfism" to "mutualism," from "aggression" to "compassion," from "exploitation to "empathy," from "isolation" to "connectedness," from "hierarchy" to "heterarchy," and above all from "conquest" and "domination" to "creativity" and "caring."[47]

There is a tendency to look back at the past and conclude that changes and transformations in history must have been adaptive. The truth is that, in the evolutionary time scale, the blood-drenched five thousand years of recorded history following the fundamental socio-cultural shift briefly traced here is far too short a period to warrant this assumption. It may or may not be true that, at a certain point in our history, a dominator system was adaptive. But we do not have sufficient evidence to decide this question.

There is, however, substantial evidence that at this point in cultural evolution a dominator system is not adaptive, that it is in fact extremely maladaptive.[48] In our highly advanced technological age, when nuclear weaponry puts in human hands the kind of total destructive power once attributed only to God, the socialization of men for conquest and domination is obviously maladaptive. If it is true that men are generally more predisposed to learn violent behaviors than women, this would be all the more urgent reason for an immediate global shift to a system where these types of behaviors are not systematically inculcated in male children through every conceivable means of socialization, from toy guns to the equation of "masculinity" with aggression and conquest.[49]

But can this global transformation, this "metamorphosis in basic cultural premises and all aspects of social relations and institutions"[50] and the accompanying "drastic changes in the norm stratum"[51] happen in time? On the negative side there is an enormous pressure for self-preservation in the system. Yet on the positive side there is an even more powerful pressure for long-term self-preservation. This is the survival impetus of a uniquely creative species that *can* change its way of thinking and acting to adapt to a world in transformation.

NOTES

1. For some examples of new approaches to evolution exploring both systems maintenance and systems change, see e.g. I. Prigogine and I. Stengers, *Order Out of Chaos*, New York: Bantam Books, 1984; N. Eldredge and S. J. Gould, "Punctuated Equilibrium: An Alternative to Phyletic Gradualism, in N. Eldredge, *Time Frames*, New York: Simon and Schuster, 1985; V. Csanyi, *General Theory of Evolution*, Budapest: Akademiai Kiado, 1982; E. Laszlo, *Evolution*, Boston: New Science Library, 1987; E.

J. Chaisson, *Cosmic Dawn*, Boston: Little, Brown, 1981. For interesting new mathematical models of systems dynamics, see R. Abraham and C. Shaw, *Dynamics: The Geometry of Behavior*, Santa Cruz: Aerial Press, 1984.

2. These models are described in detail in R. Eisler, *The Chalice and The Blade*, San Francisco: Harper & Row, 1987. See also Eisler, "Gylany: The Balanced Future," *Futures*, Vol. 13, No. 6, December 1981; R. Eisler and D. Loye, "The Failure of Liberalism: A Reassessment of Ideology from a New Feminine-Masculine Perspective," *Political Psychology*, Vol. 4,, No. 2, 1983; R. Eisler, "Beyond Feminism: The Gylan Future," *Alternative Futures*, Vol. 4, Nos. 2-3, Spring/Summer 1981.

3. An important distinction must be made between *domination* and *actualization* hierarchies. The term hierarchy in this work refers to domination hierarchies, or hierarchies based on force or the express or implied threat of force, which are characteristic of the human rank orderings in male-dominant societies. Such hierarchies are very different from the kinds of hierarchies found in progressions from lower to higher orderings of functioning, for example in the relationship between cells, organs, and the entire living organism.

4. I. Takamure, *Joshi no Rekishi*, Vol. I & II, Tokyo: Kodansha Bunko, 1975; A. Y. Carter, "Japanese Society and Women," in *Women in Asia*, Christian Women's Conference of Japan, 1979; P. Mische, "Feminism, Militarism, and the Need for an Alternative World Security System," in *Alternative Futures*, Vol. 4, Nos. 2-3, Spring/Summer 1981.

5. C. Koonz, "Mothers in the Fatherland: Women in Nazi Germany," in R. Bridenthal and C. Koonz, *Becoming Visible: Women in European History*, Boston: Houghton Mifflin, 1977; G. Mosse, editor, *International Fascism*, Beverly Hills, California: Sage, 1979; G. Mosse, *Nazi Culture*, New York: Bantam, 1966.

6. The British documentary, *Masai Women*, by C. Curling and anthropologist M. Llewelyn-Davis gives a vivid picture of Masai values and social structure. London: Granada TV, 1980.

7. F. Brenner, "Khomeini's Dream of an Islamic Republic," *Liberty*, Vol. 74, No. 4, July/August 1979.

8. R. Eisler, *The Chalice and the Blade*, San Francisco: Harper & Row, 1987. See also R. Eisler, "Human Rights: Toward An Integrated Theory for Action," *Human Rights Quarterly* Vol. 9, No. 3, August 1987.

9. *Ibid.*

10. See e.g. C. Turnbull, *The Forest People: A Study of the Pygmies of the Congo*, New York: Simon and Schuster, 1961; P. Draper, "!Kung Women: Contrasts in Sexual Egalitarianism in Foraging and Sedentary Contexts," in *Toward an Anthropology of Women*, R. Reiter, editor; New York: Monthly Review Press, 1975.

11. See e.g. J. Bachofen, *Myth, Religion, and Mother Right*, Princeton: Princeton University Press, 1861, 1967.

12. J. Mellaart, *Catal Huyuk*, New York: McGraw-Hill, 1967.

13. M. Gimbutas, *The Goddesses and Gods of Old Europe*, Berkeley: University of California Press, 1982.

14. J. Mellaart, *Catal Huyuk*, New York: McGraw-Hill, 1975.

15. M. Gimbutas, *The Goddesses and Gods of Old Europe*, Berkeley: University of California Press, 1982, 237; M. Gimbutas, "The First Wave of Eurasian Steppe Pastoralists into Copper Age Europe," *Journal of Indo-European Studies*. Vol. 5, No. 4, Winter 1977.

16. J. Mellaart, *Catal Huyuk*, New York: McGraw-Hill, 1967; M. Gimbutas, *The God-desses and Gods of Old Europe*, Berkeley: University of California Press, 1982; J. Hawkes, *Dawn of the Gods*, New York: Random House, 1968; N. Platon, *Crete*, Geneva: Nagel Publishers, 1966. See also R. Eisler, *The Chalice and The Blade*, San Francisco: Harper & Row, 1987.

17. J. Hawkes, *Dawn of the Gods*, New York: Random House 1968.

18. N. Platon, *Crete*, Geneva: Nagel Publishers, 1966, p. 177, p. 148.

19. R. Eisler, *The Blade and the Chalice: Technology at the Turning Point*, paper presented at General Assembly, World Futures Society, Washington, D.C. 1984. I am indebted to Buckminster Fuller for his ideas on this broader view of technology as well as to Lewis Mumford's work for his use of the term "technics."

20. For an incisive analysis of how even biological evolution may have been a co-creative process, see S. Washburn, "Tools and Human Evolution," *Scientific American*, Vol. 203, No. 48, September 1960.

21. See e.g. N. Tanner, *On Becoming Human*, Boston: Cambridge University Press, 1981; A. Zihlman, "Women in Evolution, Part II: Subsistence and Social Organization among Early Hominids," *Signs*, Vol. 4, No. 1, Autumn 1978; J. Lancaster, "Carrying and Sharing in Human Evolution," *Human Nature*, Vol. 1, No. 2, February 1978.

22. See e.g. C. Turnbull, *The Forest People: A Study of the Pygmies of the Congo;* New York: Simon and Schuster, 1961; P. Draper, "!Kung Women: Contrasts in Sexual Egal-itarianism in Foraging and Sedentary Contexts," in *Toward an Anthropology of Women*, R. Reiter, editor, New York: Monthly Review Press, 1975.

23. A. Leroi-Gourhan, *Prehistoire de l'Art Occidentale*, Paris: Edition D'Art Lucien Ma-zenod, 1971. The evidence for this point is detailed in R. Eisler, *The Chalice and The Blade*, San Francisco: Harper & Row, 1987.

24. J. Mellaart, *The Neolithic of the Near East*, New York: Charles Scribner's Sons, 1975.

25. See e.g. J. Mellaart, *The Neolithic of the Near East*, New York: Charles Scribner's Sons, 1975; J. Mellaart, *Catal Huyuk*, New York: McGraw-Hill, 1967; M. Gimbutas, *The Goddesses and Gods of Old Europe*, Berkeley: University of California Press, 1982; E. Neumann, *The Great Mother*, Princeton: Princeton University Press, 1955.

26. See note 2 above.

27. See e.g. M. Gimbutas, "The First Wave of Eurasian Steppe Pastoralists into Copper Age Europe," *The Journal of Indo-European Studies*, Vol. 5, No. 4, Winter 1977; J. Mellaart, *The Neolithic of the Near East*, New York: Charles Scribner's Sons, 1975; G. Childe, *The Dawn of European Civilization*, New York: Random House, 1964.

28. E. James, *The Cult of the Mother Goddess*, London: Thames and Hudson, 1959.

29. M. Gimbutas, *The Goddesses and Gods of Old Europe*, Berkeley: University of Cali-fornia Press, 1982; M. Gimbutas, "The First Wave of Eurasian Steppe Pastoralists into Copper Age Europe," *Journal of Indo-European Studies*, Vol. 5, No. 4, Winter 1977.

30. E. Pagels, *The Gnostic Gospels*, New York: Random House, 1979. This theme is developed in R. Eisler, *The Chalice and The Blade*, San Francisco: Harper & Row, 1987.

31. *Ibid.*

32. *Ibid.*

33. See e.g. N. Platon, *Crete*, Geneva: Nagel Publishers, 1966.

34. For an important work on how both biological and social systems replicate themselves, see V. Csanyi, *General Theory of Evolution*, Budapest: Akademiai Kiado, 1982.

35. T. Roszak, "The Hard and the Soft," in *Masculine/Feminine*, B. Roszak and T. Roszak, editors, New York: Harper Colophon Books, 1969; J. Nash, "The Aztecs and the Ideology of Male Dominance," *Signs, Vol. 4, No. 2, Winter 1978;* D. McClelland, *Power: The Inner Experience*, New York: Irvington, 1975; D. Winter, *The Power Motive*, New York: The Free Press, 1973. These neglected historical patterns are detailed in R. Eisler, *The Chalice and The Blade*, San Francisco: Harper & Row, 1987.

36. T. Roszak, "The Hard and the Soft," in *Masculine/Feminine*, B. Roszak and T. Roszak, editors, New York: Harper Colophon Books, 1969.

37. J. Nash, "The Aztecs and the Ideology of Male Dominance," *Signs*, Vol. 4, No. 2, Winter 1978.

38. D. McClelland, *Power: The Inner Experience*, New York: Irvington, 1975.

39. D. Winter, *The Power Motive*, New York: The Free Press, 1973.

40. S. Rowbotham, *Women, Resistance, and Revolution*, New York: Vintage, 1974. These alternations between the "attractors" of a partnership and dominator model are examined in R. Eisler, *The Chalice and The Blade*, San Francisco: Harper & Row, 1987.

41. See e.g. R. Eisler, "Human Rights: The Unfinished Struggle," *International Journal of Women's Studies*, Vol. 6, No. 4, September/October 1983.

42. See e.g. I. Prigogine and I. Stengers, *Order Out of Chaos*, New York: Bantam, 1984 and D. Loye, *The Psychology of Prediction*, work in progress.

43. G. R. Taylor, *Sex In History*, New York: Ballantine, 1954.

44. See e.g. Wooley quoted in Hawkes, *Dawn of the Gods*, New York: Random House, 1968, p. 73.

45. M. Mesarovic and E. Pestel, *Mankind at the Turning Point*, New York: Dutton, 1974, p. 157.

46. *Ibid.*

47. *Ibid.* See also e.g. W. Harman, "The Coming Transformation," *The Futurist*, February 1977; J. Salk, *Anatomy of Reality*, New York: Columbia University Press; H. Henderson, *The Politics of the Solar Age*, New York: Anchor Books, 1981.

48. See e.g. R. Eisler, "Violence and Male Dominance: The Ticking Time Bomb," *Humanities in Society*, Vol. 7, Nos. 1 and 2, Winter/Spring 1984.

49. For some works on how human behavior is not genetically preprogrammed, but rather the product of a complex interaction between biological and social/environmental factors, see e.g. R. Hinde, *Biological Bases of Human Social Behavior*, New York: McGraw-Hill, 1974; H. Lambert, "Biology and Equality: A perspective on Sex Differences," *Signs*, Vol. 4, No. 1, Autumn 1978, pp. 97-117; R. Eisler and V. Csanyi, *Human Biology and Social Structure*, work in progress. For two excellent works exploding the myth that "man is a born killer," see A. Montagu, *The Nature of Human Aggression*, New York: Oxford University Press, 1976 and R. C. Lewontin, S. Rose, and L. J. Kamin, *Not In Our Genes*, New York: Pantheon, 1984.

50. W. Harman, "The Coming Transformation," *The Futurist*, February 1977.

51. M. Mesarovic and E. Pestel, *Mankind at the Turning Point*, New York: Dutton, 1974.

Note: I also want to acknowledge the help of my friend and colleague, Barbara Honneger, with the task of naming the phase-changes described in this paper.

Subject and Name Index